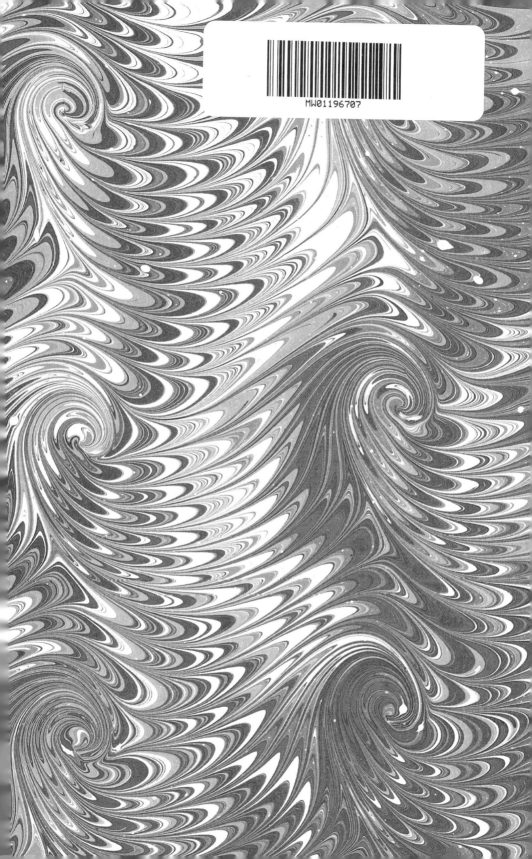

WRONG PLACE! WRONG TIME!
George C. Kuhl

Fort Benning, Georgia
16 July 1994

To Brian Davis —

Thanks for your interest in this story and for adding it to your military library.

A great many fine aircrew members helped me put it all together.

Best Regards —
George C Kuhl

*The 305TH Bomb Group &
the 2ND Schweinfurt Raid
October 14, 1943*

WRONG PLACE! WRONG TIME!

George C. Kuhl

Schiffer Military/Aviation History
Atglen, PA

Front cover photo courtesy of Harry J. Task

Book Design by Robert Biondi

First Edition
Copyright © 1993 by George C. Kuhl.
Library of Congress Catalog Number: 92-62388

All rights reserved. No part of this work may be reproduced or used in any forms or by any means – graphic, electronic or mechanical, including photocopying or information storage and retrieval systems – without written permission from the copyright holder.

Printed in the United States of America.
ISBN: 0-88740-445-6

We are interested in hearing from authors with book ideas on related topics.

Published by Schiffer Publishing Ltd.
77 Lower Valley Road
Atglen, PA 19310
Please write for a free catalog.
This book may be purchased from the publisher.
Please include $2.95 postage.
Try your bookstore first.

Preface

On 13 December 1944, late in the evening, the Kuhl crew arrived at the 305th Bombardment Group (Heavy), which was located at Chelveston, England. This was some 14 months after the raid to Schweinfurt, Germany on 14 October 1943.

When we arrived, those who flew that day in October of the previous year were now gone, but the many permanent party personnel, those who maintained the base and aircraft, were still around. Invariably the subject of exactly what happened to the 305th that day would come up in many conversations. Even after over a year had gone by, people were still talking about the "Second Schweinfurt."

Many historians consider it the world's greatest one-day air battle of World War II. Almost 50 years later, it is still discussed and remembered.

I had often wondered what happened to all those flyers who took off that day and never returned. In September, 1988, almost 45 years later, I decided to try and find out. So this is their story.

Three tools were found that helped determine the times, positions, and flight routes flown to and from Schweinfurt. The first two were copies of the lead navigators' logs from both the 305th and 92nd Bomb Groups. The third device, a Lambert Conformal Conic Projection, Operational Navigation Chart, ONC E2, scale of 1:1,000,000, was used to plot the coordinates of these two units throughout the mission.

The military clock is used. For those not familiar with it – if it is 1440 hours subtract 1200 from it, and it is 2:40 p.m. civilian time. Before noon is just as it reads – 0910 is 9:10 a.m.

For easier reading, the Roman numerals designating military units and military places have been written out and have been replaced by Arabic numbers; e.g., VIII Bomber Command, is Eighth Bomber Command, and Stalag XVIIB is Stalag 17B.

Contents

Preface ... 5
Acknowledgments ... 8

Book I: The Mission

Chapter 1	The Eighth Air Force	15
Chapter 2	The Aircraft and Its Crew	21
Chapter 3	The Briefing	29
Chapter 4	Take Off and Assembly	41
Chapter 5	Across the Channel	55
Chapter 6	Into Holland	65
Chapter 7	At the Border	73
Chapter 8	Into Germany	83
Chapter 9	Approaching the Rhine	92
Chapter 10	The Bomb Run	105
Chapter 11	The Return Home	121

Book II: Aftermath

Chapter 12	The Next Day	133
Chapter 13	The First Twelve Minutes of Battle	145
Chapter 14	The Next Four Minutes	153
Chapter 15	The Following Eleven Minutes	160
Chapter 16	Into Burg Adendorf	169
Chapter 17	Then There Were Five	175
Chapter 18	The Last Three	187
Chapter 19	The Internees	195
Chapter 20	The Escapees	213
Chapter 21	The Evaders	221
Chapter 22	The Turning Point	247
Chapter 23	Epilogue	251

Appendixcs
    Appendix A. List of Returning Crews ... 254
    Appendix B. List of Lost Crews ... 255
    Appendix C. In Memoriam ... 261

Notes ... 266
Bibliography ... 297
Index ... 298

Acknowledgements

I would like to thank, and convey my sincerest appreciation to the many people and organizations, listed below, who all had a hand in putting together this story:

First, to the 305th Bombardment Group (H) Memorial Association roster – where it all began.

From the United States Air Force Historical Research Agency at Maxwell Air Force Base, Alabama: James H. Kitchens, III, Ph. D., Harry R. Fletcher, Historian (now retired), MSgt Gary L. McDaniel, Historian Technician

From the Washington National Records Center, Suitland, Maryland: Richard Boylan, Archivist, Kevin Abing, Archivist, Deirdre Cleary

From the Casualty and Memorial Affairs Operations Center, Alexandria, Virginia: John F. Manning, Chief, Mortuary Affairs Branch, Robert M. Long, Mortuary Affairs

From the American Battle Monuments Commission, Washington, D.C.: Martha Sell, Chief, Operations Division, Freda McGranor, Administrative Assistant

From St. Louis, Missouri: The National Personnel Records Center

From the Air Force Inspection and Safety Center, Norton Air Force Base, California: John Adkins, Chief, Flight Records (now retired), Judy A. Schladen, Flight Records Technician

From overseas: Ron W.M.A. Putz, Heerlen, The Netherlands, for his contributions of maps, photographs from the Netherlands American Cemetery and Memorial, Margraten, Netherlands, documents, and other information pertaining to the Murdock and McDarby crews.

Jean-Pierre Wilhelm, "SEQUOIA", Chemin Merox, Switzerland, for the information, maps, and numerous photographs pertaining to the Dienhart Crew.

Leo Zeuren, Baexem, Holland for his information, documents, and maps pertaining to the Willis crew.

Felix Freiherr von Loë, Burg Adendorf, Germany, for his information, documents, and photographs pertaining to the Skerry crew, and the castle at Burg Adendorf.

Dr. Alfred Hiller, Vienna, Austria.

Bundesarchiv-Militararchive RL 5/147, Germany.

From Memphis, Tennessee: Fred H. Tate, 305th Bomb Group, Frank G. Donofrio, *Memphis Belle* Memorial Association, Bill Stoots, Senior CAM/AAFTC, and Terry A. Knight, Confederate Air Force, both formally of the *Memphis Belle* Restoration Association, who provided B-17 technical information

Also, many thanks to former 305th Bomb Group members: Charles Sackerson, Delmar E. Wilson, Major General, USAF (Ret), Thomas K. McGeHee, Lieutenant General, USAF (Ret), Eugene H. Nix and Clarence R. Hall for the use of their pictures.

In addition, thanks to: Eva Wagner, Irvin Industries, Canada, LTD, Barbara Augustus, Office of Air Force History, Bolling Air Force Base, Washington, D.C., Barbara B. Valentine, reference librarian, University of Georgia, Athens, Georgia, Jean Geist, library associate, Bowling Green State University Library, Bowling Green, Ohio, Pat Birchenall, Assistant Librarian, Augusta College, Augusta, Georgia, Franklin Gibson, Barnesville, Georgia, for his assistance in locating the grave site of Douglas L. Murdock.

Acknowledgements

And my deepest appreciation for the response from those of you, from the 305th, who flew that day, and really made this story fly. Thanks for your interviews, the questionnaires you returned, the copies of diaries and logs, articles, documents, photographs, comments you provided, and the many telephone calls you made. In alphabetical order, you are:

Stanley Alukonis
Raymond C. Baus
Charles E. Blackwell
Brunson W. Bolin
Louis Bridda
Carl J. Brunswick
Roy A. Burton*
Clinton A. Bush
Alfred C. Chalker**
Joseph E. Chely
John A. Cole
Edward W. Dienhart
Gerald B. Eakle
Frederick B. Farrell
Kenneth R. Fenn
Loren M. Fink
William C. Frierson
Roger J. Goddard
Charles J. Groeninger
Robert Guarini
Bel Guber (widow of Max)
John L. Gudiatis
Frederick E. Helmick
William C. Heritage
Homer L. Hocker
Stanley J. Jarosynski
Joseph W. Kane
Ellsworth H. Kenyon
Alden C. Kincaid

Joseph K. Kocher
Steve Krawczynski***
Lester J. Levy
Richard W. Lewis
John C. Lindquist
Arthur E. Linrud
Charles B. Lozenski
Walter E. Lutz
Dennis J. McDarby
William B. Menzies
Herman E. Molen
Joseph Pellegrini
LeRoy V. Pikelis
John P. Raines
Wayne D. Rowlett
Bernard Segal
Marvin D. Shaull
Robert A. Skerry****
Jayson C. Smart
Edwin L. Smith
Alfredo A. Spadafora
Harry J. Task
John C. Tew, Jr.
Christy Zullo

* Roy A. Burton passed away on 28 July 1990
** Alfred C. Chalker passed away on 18 July 1991
*** Steve Krawczynski passed away on 22 February 1991
**** Robert A. Skerry passed away on 22 February 1989

Special thanks to Robert E. O'Hearn, Historian of the Second Schweinfurt Memorial Association, for the use of some 305th Bomb Group information from his book, *In My Book You're All Heroes.*

And finally to the following editors who provided so much guidance and encouragement: Dr. Donald R. Swanson, Wright State University Press, Dayton, Ohio; Mary L. Suggs, Editor, Stackpole Books, Harrisburg, Pennsylvania; Wm. Jerome Crouch, Editor-in-chief, The University Press of Kentucky, Lexington, Kentucky.

Dedicated to the aircrews of the 305th Bombardment Group (Heavy) who, on 14 October 1943, found themselves in the wrong place at the wrong time.

BOOK I
THE MISSION

Chapter 1

The Eighth Air Force

On 9 December 1942, just over one year after the United States entered World War II, a select group of Americans, both military and civilian, began to put together an Army Air Force list of target priorities for the destruction of the German war machine. This group was known as the Committee of Operations Analysts (COA). On 8 March 1943, the COA report was completed and submitted to the Commanding General of the Army Air Forces, General Henry H. Arnold. After his review, on 23 March 1943, the slate of targets was sent to Headquarters, Eighth Air Force and appropriate British commands and agencies for their review and comments.[1] Having already been at war with Germany since 1939, the British had compiled their index of preferred targets some years before. The American and British Combined Chiefs of Staff (CCS), representing their respective governments, agreed in principle to a decisive course of action for the strategic aerial bombardment of Germany. It would be done from England with the American heavy bombers attacking by day, and British Royal Air Force bombers striking by night. It was referred to as the Combined Bomber Offensive and was approved on 18 May 1943 by the CCS.[2]

The two major combat units of the Eighth Air Force would take the war to the Germans. While Eighth Bomber Command would carry the payload, Eighth Fighter Command would fly escort.

The Eighth Bomber Command Dilemma

The architects who put together this enormous and very crucial plan warned those at the top of the possibility the Luftwaffe (German Air Force) could well impede the success of the American bombing effort. The German government was quite aware of just how open to attack and damage its war making plants and factories were during the day. To protect these most important industrial resources from the American planes, the German fighter force had been constantly expanding its numbers and improving its tactics.[3]

To further compound this problem of the increasing enemy fighter population, the Army Air Forces did not have a long-range fighter available and operational to escort the B-17 and B-24 heavy bombers of Eighth Bomber Command all the way to the target and back on deep penetrations. Up until December 1943, Republic P-47 Thunderbolt single-engine fighters from Eighth Fighter Command were mainly used. By adding external fuel tanks their range was increased somewhat.[4] However, on long-range missions far into Germany, the bombers went on by themselves when the P-47s approached their point of no return.

At this time the friendly fighters, affectionately dubbed "little friends" by the bomber crews, reached maximum outbound range with their available fuel. They now had to leave the bombers and do a 180 degree turn so their remaining petrol would bring them safely home.

This was the period of time when the "heavies" were most vulnerable. Their only defense against the German fighter attacks was a compact formation where the maximum number of machine guns could be effectively massed against the on-rushing enemy. Once the tight bomber formations became broken and scattered, the Americans took heavy losses of men and machines. While they were trying to get the job done, they needed their "little friends" around them to keep the German fighters off their backs.

In 1942 and 1943, there were some officers in the Eighth Air Force who felt the bombers could do it all by themselves. However, when the losses began piling up from the unescorted missions, they changed their minds. They quickly joined the "Eighth" consensus; i.e., a long-range fighter was needed and as soon as possible. If not, American daylight bombing over Germany would not succeed.[5]

The British, opting to bomb at night, did not have a need for long-range fighter escort. They did not fly in formation, made individual bomb runs, and relied heavily on darkness to screen themselves from the enemy.

Despite agreement among the Americans that a long-range escort fighter was a "must", planners and decision makers continued to send the B-17s and B-24s, each with a ten-man crew, deeper and deeper into Germany. The further they went, the more losses they took. For example, on 13 June 1943, 26 ships were lost from 102 that attacked Bremen, and out of 49 aircraft that hit Kassel on 28 July, 22 failed to return.[6]

On 17 August 1943, the Eighth Air Force decided it was time to go for the ball bearing plants located in eastern Germany adjacent to the city of Schweinfurt. It was also resolved to make this a two-pronged attack into Germany by sending a second force of bombers some 110 miles farther to the southeast to bomb the Messerschmitt fighter assembly plant at Regensburg.

P-47 fighters carrying additional fuel in external belly tanks escorted the two forces as far as possible. This was in the vicinity of Aachen, Germany, where the German, Belgian, and Dutch borders all meet.

At Schweinfurt, the bombing results were officially listed by the Americans as very good.[7] At Regensburg, the bombing was considered very accurate, and the results excellent.[8] The Luftwaffe also had some very good, very accurate, and excellent claims to make. The Schweinfurt force lost 36 B-17s out of an attacking force of 183, while the Regensburg air task force dropped 24 out of a total of 126 assaulting aircraft.[9] This total of 60 heavy bomber combat losses in one day's operation was a devastating blow, and it did not include planes that received battle damage and returned to home base with dead and wounded on board.

During the latter part of summer and early fall of 1943, the Eighth Air Force continued to sandwich deep, partially escorted raids to Germany in between shorter missions to France, Belgium, and Holland. These brief trips had the friendly fighters along for the entire journey. On these occasions, it was the German fighter pilots who found it extremely difficult to attack the bombers while trying to cover their tails from the P-47s. When the bombers and their "little friends" spent the complete day working together, four-engine airplane losses were minimal while the Luftwaffe took most of the lumps.

On 6 September 1943, weather hindered bombing the primary targets, so 262 heavy aircraft, *without escort*, attacked targets of opportunity deep in Germany. They lost 45. The next day 185 of them attacked short range targets in Belgium, Holland, and France. The P-47s *stayed* with their "larger friends" the entire day, and all 185 of the attackers returned safely to England.[10]

The Eighth Air Force's next big push in bombing of German targets with limited long-range fighter escort came on 8, 9, and 10 October 1943. On 8 October, 357 planes bombed the Bremen and the Vegesack areas, and 30 ships failed to return. They lost 28 out of 352 on 9 October attacking four different targets, and on 10 October, Munster was assaulted by 236 aircraft. Enemy actions destroyed 30 of them.[11]

It was becoming apparent to many people, especially the aircrews, Eighth Bomber Command could not continue these high losses of men and machines and still maintain the offensive.

In early October 1943, an entry in the 381st Bombardment (Bomb) Group medical detachment log read, "The mental attitude on morale of the crews is the lowest that has been yet observed!" Things were so bad at the 381st, a portion of the briefing for the 14 October 1943 raid would be deleted. The medical log entry for that day read, "Crews were briefed at 0700 hours and the target was the ball-bearing works at Schweinfurt, Germany. The mention of the word 'Schweinfurt' shocked the crews completely. It will be recalled that on 17 August 1943 this group lost so heavily on this same target. Also conspicuous by its omission was the estimated number of enemy fighters based along the route. Upon checking with S-2 late, it was found that this omission was intentional and that the entire German fighter force of 1100 fighter aircraft was based within 65 miles of the course. The implications are obvious. As I went around to the crews to check our equipment, sandwiches, coffee, etc. the crews were scared, and it was obvious that many doubted that they would return . . ."[12]

After the Munster raid on 10 October 1943, Staff Sergeant Charles J. Groeninger, a gunner in the 305th Bomb Group, recalled his friend and roommate, Staff Sergeant Alan B. Citron, writing a letter home to his parents saying, among other things, ". . . I will never make it home, everyone is being shot down!" The letter was intercepted by the base censor, and needless to say portions of it were deleted.[13] Despite his lack of confidence in the way the generals were managing the limited assets of Eighth Bomber Command, Alan Citron, like so many hundreds and hundreds of others, continued to fly his missions until time ran out for him!

Bomber crew personnel were supposed to do 25 combat missions, and then rotate home. During this merry-go-round in late summer and early fall of 1943, few of them were able to find the brass ring with that magic number written on it. They were either being killed, wounded, or shot down and put into German prisoner of war camps.

Nevertheless, at the top, the unrealistic Eighth Air Force strategy and blueprint for victory remained unchanged. Those in charge seemed to have forgotten just three years earlier the Royal Air Force had broken the back of the German bomber force. On many occasions, the bombers had insisted on attacking the British homeland without their fighters flying cover for them.

As Richard W. Lewis, a member of the Kenyon crew, who flew with the 305th Bomb Group and was there, so aptly summed it up, "We should have learned from German bomber limitations in the Battle of Britain. To be unescorted was a mistake!"[14]

Recent bomber losses with part time fighter escort totaled 193: 17 August, 60; 6 September, 45; 8, 9, 10 October, 88. The "hand was writing on the wall" and for some reason the generals in command could not seem to read the message. Perhaps they chose to ignore it. Whatever their reasoning, those making the decisions had thrown away the "book" on planning a successful military operation deep into Germany. The very critical ingredient of supporting firepower, the long-range fighter being in place, which would insure acceptable bomber losses was always lacking.

In the early evening of 13 October 1943, the combat units of all three bombardment divisions of Eighth Bomber Command, including the 305th Bomb Group based at Chelveston, England, were alerted to begin bomb loading for the next day's mission. Eighth Bomber Command, with the green light from Eighth Air Force, had now concluded a second effort was needed on the ball bearing works at Schweinfurt, Germany. This decision would put the bomber crews without escort for close to three hours and fifteen minutes. This inept strategy by the Eighth Air Force leaders to force another deep thrust into Germany with extremely limited fighter escort would cost Eighth Bomber Command dearly.

During the mission, the 1st Bombardment Division received the brunt of the German Air Force's wrath for two reasons: one, record numbers of enemy fighters were committed to the battle, and two, mistakes, coupled with unwise decisions, and blunders made by several air commanders flying within the 1st Division drew the undivided attention of the attacking enemy. A unit of this division, the 305th Bomb Group, suffered the most and its misfortunes are described in depth throughout this story.

This, then, is the true story of what happened during World War II to Eighth Bomber Command, specifically the 1st Bombardment Division and the 305th Bomb Group, on the way to and from Schweinfurt on 14 October 1943 – together with the aftermath that extended until the war's end.

Officially this raid into Germany was designated Eighth Air Force Mission Number 115. It is better known as the, "Second Schweinfurt."

It was a disaster!

Chapter 2

The Aircraft and Its Crew

In October 1943, Eighth Bomber Command consisted of three divisions of heavy bombers. The 1st and 3rd Divisions were assigned the B-17 "Flying Fortress", while the 2nd Division was allocated the B-24 "Liberator."[1]

As part of the 1st Division, the 305th Bombardment Group flew the "Fortress." The tragedy that befell this and other B-17 groups on 14 October 1943 could not be blamed on the machines that took them to war. For the time frame, the B-17 was the state of the art in heavy bombers.[2]

The B-17 was developed by the Boeing Aircraft Company beginning in 1934, and throughout the years was modified a number of times to increase its capabilities. The last operational combat model was the B-17G which featured a chin turret installed beneath the Plexiglas nose. This last model had a fuselage that measured almost 75 feet in length, while the wing span was just a few inches shy of 104 feet across.[3] The "G" began arriving in England in early fall of 1943. The Edward W. Dienhart crew, which was later reassigned to the 305th Bomb Group, delivered the first B-17G to Prestwick, Scotland on 4 September 1943.[4] The 305th flew five of them on the Schweinfurt raid. The other 13 ships put up by the Chelveston outfit that day were the older "F" models.[5]

With the exception of the pilot and copilot, everyone on this ten-man crew was a gunner and the B-17F and B-17G bristled with machine guns – the "F" had ten, and the "G" thirteen. These .50 caliber weapons were best put to use when the bombers flew in tight formation. This enabled the maximum number of guns to concentrate on an attacker. This method of defense was not only highly recommended for survival, it was a *must*!

The bomber crews used the clock system and the words "low", "high", and "level" for locating objects such as enemy aircraft. The aircrews started at the nose which was 12 o'clock and went clockwise in a circle. Off the right wing was three o'clock, the tail was six, and "Nine o'clock high," simply meant, "High off your left wing."

The original Lang crew at Dyersburg, Tennessee prior to joining the 305th Bomb Group at Chelveston, England. Back row, left to right: Steve Krawczynski, tail gunner; Reuben B. Almquist, left waist gunner; Howard J. Keenan, waist gunner; Eugene Kosinski, engineer and top turret gunner; Warren E. McDonnell, radio operator; Kenneth A. Maynard, ball turret gunner. Front row, left to right: James G. Adcox, bombardier; John C. (Jack) Tew, navigator; Russel Brook, copilot; Robert S. Lang, pilot. (Photo courtesy Jack Tew, Canada)

The pilot and copilot sat up front on the flight deck or cockpit area. As you looked forward, the pilot always occupied the left seat, while the copilot was on the right side. The cockpit contained all the switches, dials, buttons, knobs, and controls to fly the aircraft.

These two pilots had four Wright Cyclone engines as the prime movers for their plane. Each engine developed a thrust of up to 1200 horsepower and utilized a three-bladed propeller with a "feathering" capability. With the "feathering" device, the drag and vibration caused by a disabled engine could be reduced to a minimum. The engine was simply shut down, and the propeller blades were turned approximately 90 degrees into the relative wind. This prevented the blades from "windmilling" which caused the aforementioned problems.[6]

The B-17 was not fast. The pilots generally climbed and cruised the ship at 150 miles per hour (mph) indicated airspeed (IAS), while the letdown was accomplished at 170 mph IAS.[7]

When carrying a full load of fuel (2780 gallons of gasoline) and bombs (6000 pounds) this airplane was slow to respond. Therefore, the formation leader always had to be extremely concerned with those that followed. Any rapid changes in acceleration, deceleration, or abrupt turns, and extended maneuvers could disrupt the formation and scatter those behind him. Gentle changes in power settings and slow, shallow turns were necessary. At altitude with this weight and using superchargers, usually a climb of 100 to 200 feet per minute was the norm. If for some reason a plane fell behind the rest of the formation, this gradual ascension made that B-17 extremely vulnerable. The ship's heavy burden, coupled with everyone around him using near maximum climbing power, could mean as much as 20 to 30 minutes for a pilot just to make up 100-200 yards to regain his lost defensive slot in the formation. Occasionally a B-17 could not catch up and became a straggler. It was a most lonesome and frustrating feeling for crews being a "single." These strays were too slow to run away and invariably drew a large crowd of unsympathetic German fighter pilots, as there was no place to hide.

In front and below the flight deck was the bombardier-navigator's area. Located on the floor between the pilots' seats was a removable, folding, two-piece, plywood door that led down to this most forward station. Upon entering this area, a person could stand up and observe the seated navigator on the right with a built in map case and drift meter plus his charts and log spread out on a small table. Everything was in easy reach.

Just up ahead, the bombardier had his bombing instruments on a panel to his left. Most of the time he used a small chair when making the bomb run as he sat directly behind and over the Norden bombsight. This extremely accurate sighting device was located at the most forward point in the aircraft.

Both the navigator and bombardier doubled as gunners. Each had machine guns with which to defend the front of the plane. Also in this section, a nose hatch on the left side of the fuselage was provided for entry and exit. In the event the bomb bay doors were not open, the pilot, copilot, and flight engineer could also use the nose hatch if they had to bail out.

The "upper local" or top turret was installed just behind and above the two pilots' seats. This "top gun" position was home for the engineer. He stood on a steel platform inside the turret as he covered the upper portion of the plane with a pair of power driven, .50 caliber machine guns. These weapons rotated 360 degrees horizontally and upwards of 85 degrees.[8] When not occupying the turret, he stood or sometimes sat in a small jump seat between the two pilots. From here, he assisted in monitoring the numerous gauges on the instrument panel and called out airspeeds during take offs and landings. Also from his two positions, the engineer could easily observe the engines and both large left and right wings that housed the self-sealing fuel cells.

The lift provided by these wings was quite evident whenever the plane was forced to fly when some "horses" were not available. Due to a fine glide ratio (i.e., for every foot of descent the aircraft moved forward so many feet) provided by these enormous airfoils, the plane had no problem holding altitude with one engine inoperative. With two engines feathered it could still hold altitude providing, of course, the crew got rid of anything that was not tied down. There was no way to lighten the load by dumping fuel, so everything else had to go.

Aft and just behind the engineer's turret was a plywood door that led directly into the bomb bay. This portal was normally kept closed to keep out the cold air. An eight and a half inch wide catwalk, supported by offset steel beams, extended rearward six and a half feet to a similar door. This narrow walkway only provided about 15 inches of chest space for a person to move back and forth, and the upright clearance was just over five feet. These dimensions necessitated a crouched, sideways travel back and forth through this low, narrow corridor.[9] Waist high rope lines extended fore and aft on both sides along this cramped passage. These cords provided support for those who were required to walk this area. The bombs hung from shackles located on both sides of the catwalk. Squeezing along this constricted path was extremely "exciting" for the engineer whenever the bombs did not release and remained in place. He had to leave his top turret, enter this cold, breezy area without his parachute, negotiate the wide open bomb bay doors, and literally kick the bombs loose while hanging on for dear life.

At the aft end of the bomb bay was this other plywood door which led into the radio compartment. This room was crowded with communications equipment. It also contained a small table and swivel chair for the radio operator, who also doubled as a gunner. A single, hand held .50 caliber weapon was mounted on two tracks in the ceiling pointing rearward. To activate, a Plexiglas ceiling panel was removed creating an aperture, and the gun was slid rearward, upward, and out the top of the fuselage. From here the radioman could protect the upper, rear portion of the aircraft. This aperture was also used by the crew as an emergency escape hatch if the ship was crash landed or ditched.

In the center of the floor of the radio compartment was a recessed storage area covered by a trap door. This repository was two feet deep, two feet wide and four feet long, and was used for storing oily rags, miscellaneous aircraft equipment, together with lusty French novels and photographs. Two extra chairs were sometimes added in the radio room for the comfort of the waist gunners when the crew flew a "milk run." This was an extremely rare mission where there was absolutely no enemy opposition. During these flights, the crew would busy itself with required military reading, taking combat photographs of each other, checking the stock market, and cleaning weapons. The only real danger on this trip was getting shot in the foot.

A second plywood door led from the radio area rearward into the waist section of the bomber. The ball turret was located just a few feet past this door. This station nestled half in and half out of the floor of the ship. To qualify for this position, the individual had to be a rather short, slim person. The gunner entered this small, fish bowl shaped turret from inside the airplane, and it took a bit of doing. This enclosure had to be hand cranked into a position where the guns, the electrical powered twin "fifties", were pointing straight downward. This exposed a panel which was 21 inches wide by 21 inches long. Generally the smallest member of the crew passed through this tiny portal and entered his battle station. He then assumed a sitting position, and once strapped into place he could traverse his weapons 360 degrees and almost 90 degrees vertically. He was responsible for protecting the underside of the B-17. Due to his cramped quarters, his parachute had to remain in the waist section of the ship. In an emergency, he sometimes needed help from his comrades to exit the turret. This made him the most vulnerable member of the team.

Eight feet behind the ball turret were the two waist gunner positions, one on either side. Prior to the arrival of the "G" model, the left and right waist windows opened inward, then slid forward and up against the inside of the fuselage. Each gunner would then move his single, post mounted, hand held .50 caliber machine gun into the aperture for firing purposes. With the usual freezing temperatures at high altitude, sometimes as low as 60 degrees below zero, these windows were always kept closed when the weapons were not needed. When the fighting was taking place, this was the coldest assignment in Eighth Bomber Command. Later in the "G" model, the gun was mounted in a stationery, Plexiglas window, and the cold air was kept out.

As one continued aft, on the left, six feet past the waist gunners' positions was the waist door. This hatch served a twofold purpose. Besides being the main entrance to the ship, it was also used in emergencies as an escape exit by the radioman, ball turret operator, and both waist gunners.

In the very back of the plane was the tail gunner. He had to crawl around the tail wheel well and strut to get into his difficult position. This crewman sat on an extra large, semi-hard, canvas covered bicycle seat. This perch had no back and was without arms. It was not all that comfortable, but he made the most of it. With all the modifications to this airplane over the years, it is a wonder that some design change engineer never placed himself on that bicycle seat – if he had, it most certainly would have been improved. Tail weaponry consisted of twin mounted (no turret) .50 caliber machine guns. In an emergency, this crewman had a long way to go to reach the waist door, so he was provided with a small escape hatch just behind him and to his left.

He, just as the others on board, played an important role on this ten-man team. Besides defending the rear of the bomber from enemy attacks, the tail gunner was indispensable when the aircraft was taxiing or landing in bad weather. There was an electric torch at his station called an Aldis lamp. During inclement weather his job was to keep this light flashing to guide the ship behind.

In the inevitable, dismal English weather, poor visibility and low ceilings were the rule rather than the exception. Landing was especially hazardous when a runway was shrouded by drizzle and fog. When a returning group arrived over the home field for landing, one by one each squadron would peel off to make a rapid, descending, left turn. It had to be a tight landing circle, as each aircraft followed the blinking light in front of it. No lamp, no runway!

As the craft climbed through 10,000 feet, the crew donned oxygen masks and hooked into the main breathing system at each of their stations.[10] This sustained life for them while they performed their tasks in the high, thin air. In the "G" model there were four separate low pressure oxygen systems.[11] With the exception of the radio compartment, which had four stations, no more than three people drew air from one system. This was an extremely valuable asset, for in the event one of these systems

A formation of B-17s somewhere over Europe. The "G" in the triangle on the vertical fin of the ship indicates it is from the 305th Bomb Group. (Author's photo)

failed, it would not affect the entire crew. There were also 13 portable "walk around" bottles with at least one at each station. These containers enabled the individual to move freely about the aircraft above 10,000 feet. They were also very useful when bailing out from high altitudes. Depending upon the amount of exertion by an individual, these transferable breathing devices provided a 6-12 minute supply of oxygen, and could be recharged by a valve near the top turret position.

The B-17 was not a difficult airplane to fly, and it was very forgiving of crew mistakes. Probably the most strenuous pilot chore was a crosswind landing.[12] The ship had a huge vertical fin and rudder, and when the wind pushed against this large tail surface the nose of the aircraft would yaw into the breeze. It took most of the pilot's strength and all his expertise to maintain a straight path (crab) over the ground. He then kicked opposite rudder to straighten up the ship just as it touched down on the runway.

The ship was stable, well built, could take an incredible amount of battle damage, and still bring its crew home. One more complement for this machine – it was built in the USA by Americans.

Chapter 3

The Briefing

The bomb loading for the 14 October 1943 mission to Schweinfurt was sent to Eighth Bomber Command units the preceding day. It was received at the 305th Bomb Group from Headquarters, 1st Bombardment Division by teletype at 1818 hours on 13 October 1943.[1] The order called for all available aircraft, known as a maximum effort, to participate. Each ship was to be loaded with six 1000 pound bombs.

The field order was received at 2334 hours that same night. Flight plans and all details of the mission were now completed. Briefing for the 305th planes' crews was set for the next morning at 0720 hours. Takeoff was scheduled for 1020.[2]

Until a briefing began, the target of any raid was on a classified, "need to know" basis. As a result, the flyers were always the last ones to hear the "good news."

The wake up call came at 0500 with breakfast at the mess hall at 0600. From there, on to the briefing.

As was the case so many times, the weather was miserable. The 305th air base, home for the 364th, 365th, 366th, and 422nd Bombardment Squadrons, was shrouded by fog, and a steady, chilling drizzle seemed to be falling everywhere.

After the aircrews finished their morning meal, they headed for the briefing. Due to the unsafe flying conditions caused by the poor weather, many of the flyers figured the mission would be scrubbed. Sergeant William C. Frierson, a spare ball turret gunner on the Holt crew for today, peered in the direction of the airfield and saw nothing. Even the runways were obscured. He was sure the mission would be called off.[3]

The 18 crews scheduled for the mission began assembling and taking their seats in the large room set aside for this procedure. These 180 plus officers and noncommissioned officers crowded together in this chamber represented a diverse cross section of combat experience. For some, it would be their first time up against the enemy. Others would be trying to reach the magical number of 25 missions to complete a full tour of combat duty. Then there was the remaining cadre carrying the load by helping the newcomers and cheering on the ones who were about to finish up and go home.

Staff Sergeant Charles J. Groeninger was assigned to the Lang aircraft today. He would be on his second flight at the left waist gunner's position.[4] On Kincaid's plane, Staff Sergeant John F. Raines, top turret gunner, and Sergeant William C. Heritage, tail gunner, were both on their initial run.[5]

Pilot Lieutenant Raymond P. Bullock had a very special reason for getting his band through today without any problems. He, together with Technical Sergeant Carl J. Brunswick, engineer and top turret gunner, and both Staff Sergeants Stanley J. Jarosynski, left waist gunner, and Alden B. Curtis, tail gunner were all flying their twenty-fifth, and last combat mission.[6]

Lieutenant Frederick E. Helmick, the bombardier on the Farrell airplane, was also on his last trip, as was Staff Sergeant Herman E. Molen.[7] Molen was flying with the Eakle crew as togglier.[8]

Crew integrity in the 305th was nonexistent. The complete teams who had received overseas crew training as a unit for several months in the United States were gone. Upon reaching the combat zone, the reality of war destroyed many of the "families" who earlier back home had worked, ate, and slept together.

Combat attrition had taken its toll – as it always does. Everyone was a replacement for "the someone" who had gone ahead of him. Today, for example, a few were ill, some had been killed, while many had flown off and never returned and were now prisoners of war. Then there were the wounded who returned from combat over Europe and spent time in the hospital. Many of these Purple Heart recipients returned to flying status and found themselves assigned to different crews. They adjusted, picked up the pieces, fell in line, and continued to march. Others were simply personnel left over from aircrews who had finished their tour and had gone home. Many were from crews who had lost their pilots early.

Sergeant Lester J. Levy, a gunner, was assigned to the Murdock outfit. He was an alternate for the left waist gunner, who had a hangover from a three day pass.[9] Lieutenant Edwin L. Smith, flying today as copilot for Murdock, had his original pilot shot down on the first mission. Smith and crew had arrived on base from the USA, and it was always standard operating procedure for a new pilot to fly his first combat mission as copilot with another, more experienced team. Smith's pilot did just that, and never returned.[10] Technical Sergeant Joseph K. Kocher was a replacement on Bullock's plane for a hospitalized ball turret gunner – and so it went.[11]

The briefing began at exactly 0720 hours. For the 305th crews attending this meeting, it would be just about the only event to take place during the remainder of the day where they, nicknamed the "Can Do" bomb group, were all at the right place at the right time.

The Briefing

After a few short opening remarks, the briefing officer reached for the drawstring of the curtain covering the large European wall chart. Behind this blind would be today's flight path plotted on the map. Tension was always high at this time. The participants would sit not knowing, but wondering, where today's battle would be fought. A hush settled over the room as the fabric covering slid open and away from the wall, displaying the enormous Lambert Conic Projection. Upon it appeared a long red line, beginning in England and extending far, far into Germany! This would be their course for today's operation.

The meeting officially opened and the quiet ended when the briefing officer announced, loud and clear, "Your target for today is Schweinfurt, Germany!"

The strategic plan of the Eighth Air Force was to dispatch 351 heavy bombers to Schweinfurt. The 1st Bombardment Division was to send 149 B-17s, consisting of nine groups, divided into three combat wings. The 3rd Bombardment Division was scheduled to put up 142 B-17s, also consisting of three wings, while the 2nd Bombardment Division was called upon to provide 60 B-24s with two understrength wings flying as one.[12]

Three specific groups were permanently assigned to fly together to form a combat wing (CBW). The 305th, 306th, and 92nd always made up the 40th Combat Wing. Which outfit would fly the wing lead was rotated on a daily basis. Today, the 92nd group would lead the 40th Wing.

The 92nd was also designated to lead the 1st Division which meant it would take the point and show the way for this long procession of bombers often referred to as the "bomber stream." Thus, the 40th wing would be followed by the 1st CBW, consisting of the 91st, 381st, and 351st groups, while the 41st Combat Wing would bring up the rear with the 379th, 384th, and 303rd groups. For defensive purposes, each of the three wings would fly the combat composite box stagger formation: i.e., one group of each wing would be stacked high on the right of the lead group, while the other one flew low and on the left of its leader.[13] The bombers would form over England and then fly a direct route across the English Channel to the enemy coast.

Penetration for the 1st and 2nd Divisions was to be near the southern tip of Holland, in the area of Zeeland, with a 35 minute interval between the two units. The 3rd Division was to enter the continent 21 miles further to the south where the Dutch and Belgian borders meet at the coast.[14] From there, the bombers would proceed to the target via the planned route. The 1st Division was to bomb the target at 1424 hours followed by the 3rd Division at 1444. The 2nd Division consisting of the 60 B-24s was to finish off the target at 1459 hours. Bombing altitude was 23,000 feet for the 1st and 3rd Divisions, while the 2nd Division would bomb from 21,000 feet.[15]

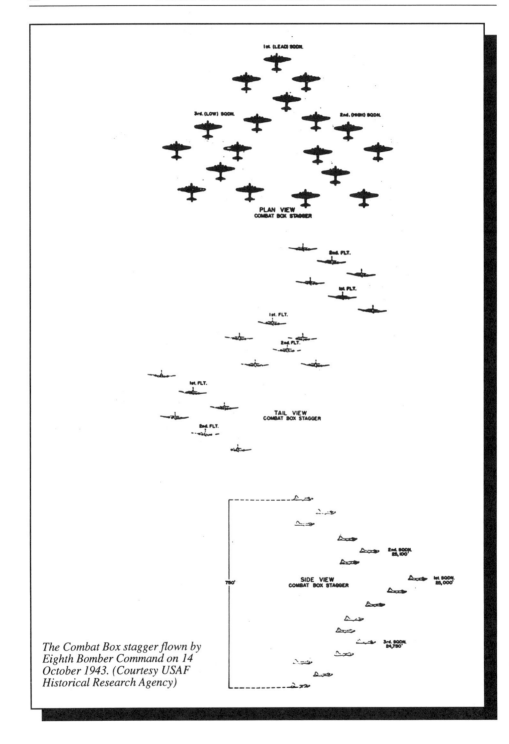

The Combat Box stagger flown by Eighth Bomber Command on 14 October 1943. (Courtesy USAF Historical Research Agency)

The American P-47 "Thunderbolt" fighter that provided escort for the B-17s and B-24s early in the war. Its armament consisted of eight .50 caliber machine guns. (Air Force magazine photo, April, 1945)

Each division was assigned one group of long-range P-47s for fighter escort protection with the 353rd Fighter Group providing air cover for the 1st Division. On the way out, the 353rd was to meet the bombers over the English Channel. These "little friends" would stay with the "heavies" as long as they could. Despite carrying an external fuel tank to give this fighter the needed "long-range", the P-47 normally only made it as far as the German border – its point of no return! It was then time to turn around and return to base. The P-47 pilots flew with these belly tanks as long as possible, but when attacked by German fighters the fuel cells had to be dropped in order to maneuver against the enemy. On the route back, the American fighters were to rendezvous with the bombers over France, just south of Rheims, and provide the convoy home. British Spitfire fighters were also to furnish limited support across the channel when the bombers departed from and returned to their English bases.[16]

The 305th Bomb Group would provide eighteen B-17s for the mission each with a ten-man crew. The group would be divided into three squadrons of six ships each. These six-ship squadrons consisted of two, three ship elements with the second element stacked directly behind and just below the lead (first) element.

The 365th Bombardment Squadron would lead with Lieutenant Joseph W. Kane, on his 23d mission, as pilot.[17] Major Charles G.Y. Normand, commanding

The British Spitfire fighter which provided limited escort for the American bombers early in the war. Armament consisted of two 20mm cannon and four .303 inch machine guns. (photo courtesy of Eugene H. Nix, Indiana)

officer of the 365th squadron, was designated as aircraft commander and group leader. He would sit opposite Kane, in the right seat, handle the radios, and have command responsibility for all group decisions while directing the 305th traffic to and from the target. Leading the 305th group was on a rotational basis, with the group commander, his executive officer, and the four squadron commanders taking turns.[18]

Major Normand had graduated from the Flying Cadet program and had been commissioned a second lieutenant on 24 March 1940. However, his experience and expertise to lead the group in combat was quite limited. Prior to joining the 305th Bomb Group on 10 July 1943, when he assumed command of the 365th Bomb Squadron, Major Normand had logged over 1600 hours of flying time since receiving his "wings."[19] Most of these hours were flying twin engine B-18As on submarine patrol. Early in the war the B-18A aircraft were mainly used for flying antisubmarine missions looking for German U-boats along the coasts of North, Central, and South America.[20] While stationed at Piarco Field, Trinidad, British West Indies in October, 1941, Normand recorded his only four-engine time. He flew twice as copilot aboard an earlier model B-17B for a total of six hours and ten minutes.

At the 305th in July, 1943, after four hours and ten minutes of B-17 pilot transition training which consisted of take offs, landings, and learning cockpit procedures, Major Normand flew his first combat flight. He logged five hours and fifteen minutes as copilot. Later that month, he flew two more combat missions as an observer, and his flying time was just over 15 hours. On 3 August, Normand received two hours and forty-five minutes of additional transitional training as a copilot. So with only the one combat flight as copilot, Normand was then signed off as an air commander to lead the 305th Group against the enemy. In mid-August, he flew two combat missions, one of them as group leader.[21]

After the August lead, Colonel Delmar E. Wilson, the 305th commanding officer, restricted Normand from leading again. He would continue to fly, but not lead. Colonel Wilson based his decision on complaints at a critique from a number of returning flying personnel who had followed Normand as he directed their August mission.[22] When Wilson was transferred to Eighth Bomber Command headquarters at the end of August, he recommended to the new 305th commanding officer, Lieutenant Colonel Thomas K. McGehee, not to let Normand lead. Wilson stated he lacked confidence in Normand and felt he was not yet qualified to command the outfit in combat.[23]

Lieutenant Colonel McGehee did not remember this conversation. He knew Normand quite well, as they had been classmates together in the Flying Cadet program. He felt Normand was competent to lead and so put him back into the group lead rotation.[24]

In September, Normand flew seven more combat missions leading the group on the 2nd and 26th of the month. On 13 October, Major Normand recorded his first combat time as pilot on a mission that was recalled due to bad weather. He flew two hours and fifty minutes. The next day, 14 October, it was his fourth turn to lead the group – this time to Schweinfurt.[25]

The briefing continued. The 366th was designated to fly high squadron, stacked high on the right of Normand's lead 365th squadron. It was to be led by Lieutenant Frederick B. Farrell. The 364th squadron, headed by Lieutenant Ellsworth H. Kenyon, would fly low squadron, stacked low on the left of Normand.

Finally, the 305th Group was to take off as desired, climb individually on course through the clouds and assemble on #40 buncher, over Thurleigh at 12,000 feet, or 1000 feet above the overcast.[26] A buncher was a designated nondirectional radio homing beacon about which the ships of a wing circled to the left until formed into three separate groups. The groups then moved on to join each other and become a wing. The three wings then gathered to make up a division, and as the three divisions assembled in a tandem arrangement, a bomber stream was formed.

After the 305th formation assembly was complete, it would continue to circle until it was time to depart the buncher and join the 92nd Bomb Group which was leading the 40th CBW. The 305th Group was to fly on the left as low group in the wing, while the 306th Group was to be stacked high on the right.[27]

The weather forecast was not too bad for the take off: broken to overcast clouds with rain, a 3000 foot ceiling, and cloud tops at 11,000 feet. Visibility would be two to three miles. For the target, the prognosis was a few scattered clouds and visibility 10 miles. On the return for landing at base, the picture was an even brighter one – a 3000 foot ceiling and visibility four to six miles. Unfortunately the meteorologist would end up with two "busted forecasts" as the weather would not perform in the morning or afternoon over England as predicted.[28]

The last portion of the briefing concerned the enemy and his intentions. The crews were now told what they could expect along their routes, out and back, from both flak and fighters.

On most of the flight path, flak would be meager and inaccurate. Over the Ruhr valley, it would be more concentrated and accurate. It would be moderate at the city of Würzburg which was the initial point (IP) for the 1st Division. It would be here, over the IP, where the bomber pilots would make their final turn, roll level, open their bomb bay doors, and begin the bomb run heading straight for the target – Schweinfurt. For the target, the intelligence briefing officer indicated antiaircraft fire would be a factor – intense, and extremely accurate.[29]

The fighter opposition estimate brought a complete hush from the entire gathering. Many of these crewmen sitting and listening today had been on the first Schweinfurt raid on 17 August 1943. It was there the most intense German fighter resistance of the war had occurred. A fierce battle had ensued, with the Eighth Bomber Command suffering staggering bomber losses. The 305th had attacked the target with 27 aircraft, and two failed to return. They had been extremely fortunate losing just the two. Two other groups did not fare as well. The 91st Bomb Group dropped 10, while the 381st Group gave up 11.[30]

Many present now had trouble following the intelligence briefing. The officer spoke slowly, his words loud and clear. Those gathered just could not believe what he was telling them. Today he stated these combat crews could see more enemy aircraft than they encountered on the last visit to Schweinfurt. They were told there were now some 1100 German fighters based within 65 miles of today's course.[31] Hopefully, a good number of them would not be operational due to repair, periodic maintenance, lack of parts, etc. However, those present were informed possibly as many as 400 unfriendly fighters would show up, and this would be 100 more than the first visit to Schweinfurt.[32]

The single-engine fighters would consist of Focke-Wulf Fw 190s, and Messerschmitt Bf 109s (Me 109), both firing 13mm machine guns and 20mm cannon. It was also predicted many different models of twin-engine aircraft would appear on the scene, such as the Messerschmitt Bf 110 (Me 110), Junkers Ju 88, and the Messerschmitt Me 210. Besides being armed with machine guns and small cannon, they would also be firing rockets into the ranks of the bombers. The enemy would come from out of the sun, head-on, directly from the tail, and occasionally from the side positions. They would attack singly and in pairs. They probably would not come at the bombers until the friendly fighter escort departed for home.[33]

The Messerschmitt Bf 109 German single-engine fighter which had three 20mm cannon and two 13mm machine guns. (Photo courtesy of Eugene H. Nix, Indiana)

The Messerschmitt Bf 110 German twin-engine fighter/bomber. Its armament included two 20mm or 30mm cannon, four 7.9mm machine guns, and two 81Z machine guns. (Photo courtesy of Eugene H. Nix, Indiana)

The Focke-Wulf Fw 190 German single-engine fighter. It carried two 13mm machine guns, and up to four 20mm cannon. (Photo courtesy of Eugene H. Nix, Indiana)

It was recommended that each aircraft carry 6800 rounds of .50 caliber ammunition, as this had been ample in the past. However, numerous gunners were making written and mental notes to load as much ammo on board as possible!

With the briefing about to end, one more thing remained to be said. The commanding general of Eighth Bomber Command, Brigadier General Frederick L. Anderson, decided to send an inspirational message to his flyers who were participating in today's venture. The message was sent on 13 October at 2315 hours through the commanding officer of the 1st Bombardment Division. It contained the following:

> "TO ALL LEADERS AND COMBAT CREWS. TO BE READ AT BRIEFING. THIS AIR OPERATION TODAY IS THE MOST IMPORTANT AIR OPERATION YET CONDUCTED IN THIS WAR. THE TARGET MUST BE DESTROYED. IT IS OF VITAL IMPORTANCE TO THE ENEMY. YOUR FRIENDS AND COMRADES THAT HAVE BEEN LOST AND THAT WILL BE LOST TODAY ARE DEPENDING ON YOU. THEIR SACRIFICE MUST NOT BE IN VAIN. GOOD LUCK, GOOD SHOOTING, AND GOOD BOMBING. [signed] ANDERSON."[34]

History does not indicate whether this message had the desired effect or not, but many who flew that day still remember the words. In General Anderson's enthusiasm to get his team "up" for this air battle, he appeared to have lacked good judgment in several areas. First, the message was originally classified "secret", but then for some unexplained reason it was decided to send it "in the clear!"[35] This must have been a pleasant surprise for the German radio intercept operators, who were constantly monitoring Allied radio transmissions. While the communication did not disclose specifics, it most certainly alerted the enemy to the fact that the 1st Division of Eighth Bomber Command was involved in something "very big" today. Secondly, the wording in sentence six of the communication did not seem very appropriate for the occasion. When a commander orders his men into battle he does not remind them of the casualties they have already taken and the ones they will suffer today. Military men do not need to be told by the "old man" they and their buddies may not get through the day. They are quite aware of this fact, as it goes with the job! Death is a very poor topic of conversation with GIs.

The general briefing officially ended and the pilots, navigators, bombardiers, and gunners gathered by positions in separate areas for more short, specific briefings.

The Briefing

As before any mission, after the meetings were over, many of the flyers met with their respective chaplains to receive the Lord's blessing. The priest provided the "Host" for the Catholics, and the Protestant Minister furnished "Communion" for the other denominations. This was considered a time of "special dispensation" by the churches.

It was now time to assemble. At bomber bases all around England crews were now piling into trucks and jeeps and heading for their aircraft.

Chapter 4

Take Off and Assembly

The aircrews of the 305th Bomb Group began arriving at their respective aircraft a little before 0845. It was now light, but a steady drizzle was falling. Fog was everywhere. The airmen had a good hour to draw equipment, load ammunition, then check and recheck their B-17 and its individual systems prior to take off.

After that, it was a good time for "chalking." The magnificent ground crews who maintained, serviced, and kept the bombers flying also provided pieces of white, blackboard chalk for the aircrews. During the lull before take off, many of the flyers would crawl into the bomb bay and on the bombs inscribe heartwarming greetings and messages to the enemy. Then some 15 to 20 minutes before "start engines" each man would be at his station and ready to go.

However, the first order of the day was a brief gathering of each crew around its pilot. To escape the rain and dampness, some aircrews crowded into the ground crew's tent which was adjacent to the plane. Others moved into the well illuminated waist section of the bomber. Newcomers to each crew were welcomed, the pilot gave everyone his message, and last minute questions were asked and answered. As the

A 305th crew prepares to board their B-17G for a mission somewhere in Germany. (Author's photo)

305th flyers gathered for this last get-together on the ground before starting the mission, there were a lot of different faces in most of the outfits. In a short span of time, the strangers were assimilated, and the team was formed. Many of these "new people" had been wounded on previous missions, went through the healing process, and were ready to go again.

This morning over on the lead Kane crew, the left waist gunner, Staff Sergeant Charles B. Lozenski, was welcomed back to his old gang. He had suffered arm and leg wounds on an earlier mission to Kiel which required 12 weeks of recovery.[1] Also, Kane had a new tail gunner. Lieutenant H.W. Luke had replaced the normal enlisted gunner for this mission.[2] A commissioned officer, who was also a pilot and qualified with the .50 caliber machine gun, always sat in the tail of the group leader's ship. This was to keep those up front in the cockpit informed, from a pilot's point of view, what was going on to the rear.

The Bullock crew met their ball turret gunner, Staff Sergeant Joseph K. Kocher, for the first time. He was back after a two week stint recovering from shell fragments in the chest and eye received over Villacoublay, France.[3]

Lieutenant Gerald B. Eakle and his bunch greeted Staff Sergeant Herman E. Molen, gunner and togglier, who had received face and head wounds while celebrating the fourth of July over Nantes, France. He was replacing Eakle's regular bombardier, Lieutenant Joseph F. Collins, Jr., who had been promoted to low squadron lead bombardier with the Kenyon crew.[4]

At one crew assembly area, the story was a bit different. An original, intact unit gathered around a new pilot. The pilot, Lieutenant Alden C. Kincaid, had already flown 21 missions with the 305th. Prior to joining the Eighth Air Force, Kincaid had been a pilot for the Royal Air Force flying Wellingtons, a British twin-engine bomber that had flown nightly raids over Germany earlier in the war. For today he had been assigned as pilot for the Norman W. Smith crew.[5]

Lieutenant Smith, together with his crew members, had just recently arrived from the United States. Al Kincaid would provide the necessary experience and leadership for them today. Lieutenant Smith would move over to the right seat and fly as copilot. Since there was no room aboard for Smith's regular copilot, he was given the day off.[6]

Smith was at the wrong end of the war. He wanted to be fighting the enemy in the Pacific Theater of Operations. Just a few years before he had been living with his family on a pineapple plantation in Hawaii. He, like many others, had gone outdoors early on Sunday morning to look up and stare at all the airplanes flying in the direction of Pearl Harbor. The Japanese had arrived to start the war in the Pacific.[7]

The Dienhart crew was intact except for their left waist gunner, who had been wounded on the previous mission. He was now replaced with Staff Sergeant Christy Zullo. After their pilot was shot down on his first mission, Zullo's regular crew had been broken up and used as fillers.[8] Staff Sergeant Raymond C. Baus, a Dienhart regular at ball turret, was back after spending several days in the hospital with a frostbitten hand.[9]

With the exception of today's navigator, Lieutenant Max Guber, and the bombardier, Lieutenant Frederick E. Helmick, the crew of Lieutenant Frederick B. (Barney) Farrell was complete.[10]

Farrell, about to fly his 20th mission, was indeed fortunate. So were his people. On their second visit to the continent, German fighters gave them a fit. The pilot's oxygen system was shot out, power was lost on one engine, and they became a straggler. Six Bf 109s were in the process of shooting their aircraft to pieces when along came four British Spitfires flown by Polish pilots. The Germans turned and fled with the Poles in hot pursuit. Lieutenant Roy A. Burton, Farrell's copilot, often wondered if those Polish pilots ever made it home that day. It appeared they were too far into Germany to have enough fuel remaining to get back to England.[11]

The Farrell crew was still deep in Germany and copilot Burton, along with right waist gunner Staff Sergeant Jayson C. Smart, had been wounded. Burton was hit in the lower right leg by cannon shell fragments. Smart, flying on his first mission, received wounds to the head and left arm also from fragments of 20mm cannon fire. The force of the explosion knocked him forward onto the upper portion of the ball turret. His hand held .50 caliber machine gun was disabled, and he realized the electrical circuit in his flying suit was inoperative. He staggered into the radio compartment, where the radio operator provided first aid for his wounds. He then helped the radio operator by feeding ammunition to the gun being fired by the radioman. Despite having gloves on and disabled electric boots, he suffered severe frostbite to his hands and feet. At the time, the temperature was hovering around minus 60 degrees at the altitude of the aircraft. He spent considerable time in the hospital and managed to avoid any amputation of his frozen fingers and toes.[12]

It was now 0950 hours, and everyone in the 305th was set to go. The poor flying weather persisted. The ragged ceiling was between 100-200 feet. The drizzle had turned to rain. The rain coupled with the fog reduced visibility from between one-quarter to one-half mile. Engines were started and checked. Taxiing began and so did the Aldis lamps. The fog became thicker. It now appeared the mission would be called off, but the airplanes continued to inch forward. It was slow going following the concrete taxiways in the murky weather. Somewhere out there was the take off runway, but it would be a while before the bombers reached it.

Each pilot eased his ship along following the blinking lamp that was flashing from the tail section of the plane in front of him. At times the only thing visible in front of the creeping ships was that flashing lamp.

Periodically everyone would come to a stop. Someone up ahead had come to a junction in the highway and was slowly executing a turn. The taxiways were not all that wide and running off the side onto the wet, muddy soft shoulders could mean disaster. A wheel buried in the mud with a full bomb and gas load on board meant a long time to dig out – maybe hours.

Major Charles G.Y. Normand, the group leader, was supposed to lead his 365th Squadron onto the runway first, followed by the high, then the low squadron. The first six ships found the runway, but they were not in the proper sequence. Normally each six-ship squadron would line up together on the runway with the leader first and his two wingmen on either side – #2 on the right and #3 on the left. The second element, led by the #4 ship, would be directly behind the lead ship with his right and left wingmen also in position (#5 and #6). As they began taking off, the next squadron would form on the runway and so on.

Due to the dismal weather, Lieutenant Joseph E. Chely, who was supposed to be fourth for take off, arrived first. The plane to be flown by Lieutenant Raymond P. Bullock followed Chely onto the runway and pulled into his designated #3 spot on Chely's left. Besides being out of sequence, the 365th proceeded to line up for take off with seven ships instead of six. Somehow the aircraft being taxied by Lieutenant Robert S. Lang, from the 366th (high squadron), now showed up on Chely's right

B-17s of the 305th Bomb Group moving past the main hangar along the taxiway to reach the take off runway. (Photo courtesy of Clarence R. Hall, South Dakota)

Take Off and Assembly

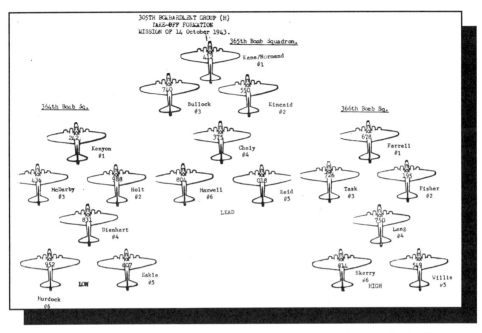

The briefed 18 ship take off formation of the 305th Bomb Group for 14 October 1943. (USAF Historical Research Agency)

side. Next, Lieutenant Joseph W. Kane, flying Major Normand's lead aircraft, moved into the #4 position. Lang was now sandwiched in between the first ship on the runway, occupied by Joe Chely, and the fourth one driven by Joe Kane. Lang had no place to go, so he stayed put. The remaining three ships of the lead squadron took up their positions on either side and behind Kane's B-17. This change in the alignment would cause some delay while assembling over Thurleigh. It meant several followers would arrive on station much earlier than the leaders. For Lieutenant Lang it would be a long wait up there until Lieutenant Barney Farrell, his squadron leader, arrived.

The planes sat for a few minutes waiting for a break in the weather. Since the visibility was so limited, each individual take off was set for a one minute interval rather than the normal 30 seconds. However, due to the constant changing visibility from rain and swirling fog, ships took off whenever they could, as they were unable to maintain the one minute take off time between ships. Lieutenant R.E. Reid, who took off sixth, had to wait nearly four minutes before he had enough visibility to follow the Kincaid aircraft down the runway. Both Lieutenant Charles W. Willis, Jr., in the high squadron, and Lieutenant Ellsworth H. Kenyon, leader of the low squadron, had to wait three minutes for the weather to improve to bare minimums before they could start down the runway and become airborne.[13]

At 1020 hours Chely was given the green light, and away he went followed shortly thereafter by Lang and Bullock.[14] Kane with group leader Normand aboard broke ground at 1026 – six minutes late.[15] Take off for the last and 18th aircraft of the 305th, flown by Lieutenant Douglas L. Murdock, of the low squadron, occurred at 1046.[16]

Just after leaving the ground it was "wheels up" and into the low overcast. The pilots would now fly using instruments as each ship immediately began a turn to a predetermined heading of 280 degrees. They continued through the clouds, on this heading, for so many minutes at 150 miles per hour indicated air speed and climbed at 300 feet per minute until they broke out on top of the bad weather.[17]

Some 23 minutes after takeoff, Chely's aircraft popped out of the overcast at 7000 feet. He leveled off, started a shallow left turn, and began circling the assembly area, keeping the needle of his radio compass pointing off his left wing at the radio beacon. He would establish a circle with an approximate three mile radius at #40 Combat Wing buncher, over Thurleigh, while waiting for the others to arrive. One by one the other aircraft began to break in the clear and attempt to find and join the circle.

At 1046 Murdock, the last 305th pilot to take off, began his take off roll. He would be flying the "Purple Heart corner" for the 364th. This was the most vulnerable position in the formation, as it was the outside, left rear slot in the low squadron. It would require him some 40 minutes to reach 12,000 feet which only gave him about seven minutes leeway before his group departed Thurleigh. Usually the last one off had from ten to twelve minutes to locate his outfit and get into position. This morning everything continued to run late.

As Murdock's plane continued moving down the runway, the ship had trouble picking up airspeed. Normally the B-17 pilots liked to reach 110-115 miles per hour before applying back pressure on the control column to leave the ground. As the ship continued to use up most of the runway, the air speed indicator had barely reached 100. As he began to near the end of the runway, Murdock eased back on the controls. The ship hesitated momentarily then slowly responded and broke ground. It was none to soon, as the end of the runway passed just beneath the ship. During an intercom check later in the morning, the copilot, Ted Smith, found out why the ship had not reach optimum take off speed. In their eagerness not to run out of .50 caliber ammunition during the mission, the crew had inadvertently overloaded the aircraft.[18]

Normand's B-17, which left the ground at 1026, cleared the overcast at 1049 hours and took the lead. The lead pilot, Lieutenant Joseph Kane, continued the left circle started by Chely and began a slow climb. The 305th had to be over Thurleigh, five miles from Chelveston, at 12,000 feet by 1132.[19] It was then to join the 92nd Bomb Group and fly the low group position. The 306th would combine with them, fly high group, and this would form the 40th Combat Wing.

Normally a group leader, while circling, continuously fired a certain colored flare; e.g., red-green, red-yellow, green-green, etc. to indicate his location to the other ships of his group who were looking for him. While gathering his own flock, Major Normand also had to be looking for the colored flares of the 92nd Group, led by Colonel Budd J. Peaslee, so he could locate his wing leader and join him on time. Timing in combat, like in most everything else people do, is essential to get the job done. The sky was filled with scattered B-17s and B-24s. With the poor visibility caused by the ragged layers of clouds many pilots were milling about trying to find their respective groups. Flares were going off all over the sky as the bombers chased their colors. It was slow going for everyone up there.

Gradually the 305th began to gather, as most of the ships found their positions in the formation. However, Normand was having problems picking out the 92nd's flares. Soon it would be time to depart the assembly area, and he still had not located his wing leader. Despite not having all his 18 ships in their designated slots, Major Normand had Lieutenant Kane continue a slow climb in order to reach 12,000 feet on time. As the 305th passed through 10,000 feet, it was now time to don oxygen masks.

The climb continued. The group assembly still was not complete. Lieutenant Dennis J. McDarby and his crew were playing "catch up" with Kenyon's low squadron, as was Murdock, the last one off.

Above and over on the right in Normand's high squadron Lieutenant Farrell, the squadron leader, was still looking for his left wingman, Lieutenant Harry J. Task. Task had been the ninth ship off, so he had been right in the middle of the pack. Yet he was nowhere to be seen. Farrell continued the 366th's circle, following Normand, and giving Task and his crew a bit more time to find him.

Farrell waited several more minutes and then decided to make some changes in his squadron formation. The empty #3 position left by Task was filled by Lang's aircraft, which moved up and left from the #4 slot. Lieutenant Robert A. Skerry, who had been flying off Lang's left wing in the #6 spot, moved right and forward and took over Lang's place. Lieutenant Charles W. Willis, Jr. slid in alongside Skerry's right wing and kept his original #5 position.

Normand's aircraft reached its initially assigned altitude of 12,000 feet, leveled off, and maintained its circle. As Normand continued to scan the area for the 92nd,

Lieutenant Jack J. Edwards, his navigator, reported three or four groups were visible flying around Thurleigh, the wing assembly area, but no flares were being fired. Normand also saw them.[20] Rather than fly towards them in order to get their identity, he chose to maintain his position a bit longer and continue circling in the hope of sighting the leader's flares.

Major Normand and Lieutenant Edwards did not realize it, but they were looking directly at the other two groups in their 40th Combat Wing, the 92nd and the 306th. The third one was the 381st from the 1st Combat Wing. Due to the weather, the 381st Group was also late and had problems finding the 91st and 351st groups which made up the 1st Combat Wing. It had joined the 40th Wing in error.[21] However, the 381st did not hang around long. It quickly headed for the "coast out" to try and catch its leader, the 91st Group led by Lieutenant Colonel Theodore Milton.

This high concentration of aircraft over England began to draw the attention of the German radar. At 1130 hours numerous radio transmissions began taking place between the enemy's radar plotting service and several fighter ground control units. One transmission indicated that all fighter groups would be at final readiness state at 1137.[22]

While circling, looking for flares, and still trying to locate the 92nd and 306th Groups, Normand began running late. Colonel Peaslee's 92nd Group minus the 305th, but with the 306th trailing 1000 feet above and behind, departed Thurleigh on time at 1132 hours at the correct altitude of 13,000 feet.[23]

The two groups were now flying the briefed southwesterly course of 247 degrees heading for Stony Stratford. They would continue to fly a prescribed route around England consisting of systematic doglegs to five different check points. As it moved from place to place, the 40th Combat Wing would pick up the remaining combat wings along the way to form the 1st Bombardment Division. The last check point would be Orfordness on the coast. With the 40th leading, the 1st Bombardment Division was to arrive at the "coast out" (Orfordness) at 1230 and be intact as it proceeded east towards the enemy.

Also at 1230 the 3rd Bombardment Division was to depart the coast at Clacton some 23 miles to the south of Orfordness. At 1305 hours, the B-24s of the 2nd Bombardment Division were to depart Orfordness, 35 minutes behind the other two divisions.

Most of the 1st Division bombers were now on top of the bad weather, but visibility was still limited by patchy clouds, drizzle and haze. At 1139 the 92nd crossed over the first check point of Stony Stratford, altitude 13,000 feet, and one minute early.[24] It was here the 306th slid into its assigned, high position on the right to form the 40th Combat Wing, still minus the 305th. The two groups then made a gradual right turn to a northwesterly heading of 325 degrees and started for Daventry, the second leg of the trip.

Two minutes later, at 1141, the 305th left Thurleigh departing *nine minutes late* on a heading of 280 degrees.[25] The Task aircraft and crew was still absent. Lieutenant Task never did locate his group. He arrived on top of the overcast, observed many different colored flares being fired, but he could not identify the ones from the 305th.[26] Due to the weather, Task could not return to Chelveston and settled for a landing at another field at West Reynham at 1345 hours.[27]

As he headed west at 12,000 feet directly for Daventry, Major Normand continued to keep strict radio silence. The 305th leader decided to bypass Stony Stratford, the first checkpoint to the southwest, to make up for lost time. Hopefully he would see his other two groups near Daventry, the second checkpoint.

For some reason Normand now deviated from published orders and slowed his indicated air speed. He would hold a steady 145 miles per hour IAS while cruising and climbing around the various check points to the coast.[28] The wing he was trying to catch would maintain 155 mph IAS while cruising to reach Daventry, then it would slow down to 150 mph while climbing until reaching Orfordness at the coast.[29]

In order to straighten out the history of what Normand actually did next, it is necessary to refer to two publications: first, "The 1st Bombardment Division, Report of Operations, dated 20 October 1943", concerning the Schweinfurt raid and secondly, to the book, *Black Thursday*, written about this same mission by Martin Caidin in 1960.

Paragraph 2a(1) of the 1st Division report reads, ". . . In the 40th Combat Wing, the 305th Group did not rendezvous with the 92nd and 306th Groups due to weather and differences in actual times at rendezvous points. The 305th Group proceeded to Daventry arriving approximately nine minutes early and failing to find the 40th Combat Wing joined the 1st Combat Wing and flew in trail of the 1st Combat Wing. . ."[30]

In *Black Thursday*, Caidin wrote as follows, "Making the best of a worsening situation, the commander of the 305th wisely proceeded, once he determined assembly with the 92nd and 306th Groups was impossible, to the next assembly area, where the radio beacon marked the position of Daventry. Here the 305th Group commander again searched in vain for the rest of his wing. Failing to find the 40th Combat Wing, he contacted the commander of the 1st Combat Wing – these B-17's were in visual contact with the 305th – and then assumed the trail position in this wing."[31]

With all due respect to both the operations section of the 1st Bombardment Division and the distinguished author Caidin, this is *not* what happened at Daventry, or when and how the 305th opted to join the 1st Combat Wing!

As shown in the 305th lead navigator's log, after flying just four minutes on his westerly 280 degree heading for Daventry, Major Normand changed his mind. At 1145, he ordered a right turn to the northwest to a heading of 322 degrees. So Normand would never reach Daventry! While still some 16 miles out, he had decided against a rendezvous there. Apparently he now planned to try and intercept the 92nd and 306th groups as they flew the briefed northeasterly leg that led from Daventry to Spalding.

The facts, as published in the 1st Division operations report and author Caidin's book concerning the 305th joining the 1st Combat Wing, are also not correct. At this time, Major Normand was *not* in radio or visual contact with anyone. The 305th was all by itself, and things would remain that way for a long time.

This is confirmed by Normand's navigator, Lieutenant Jack J. Edwards, in his written report submitted the next day to Lieutenant Colonel Thomas K. McGehee, the commanding officer of the 305th. This document indicates the 305th joined the 1st Combat Wing far out over the English Channel, thinking it was Colonel Peaslee and the 40th Combat Wing.[32]

It was 1145 hours when Normand began a climb to leave 12,000 feet and executed his turn to the northwest. Just one minute later, at 1146, the 92nd and the 306th were located three miles east of Daventry.[33] The 40th Combat Wing leader, Colonel Peaslee continued to search in vain for the missing 305th.

As the 92nd passed abeam of Daventry, it turned right to a heading of 55 degrees, and began a climb from 13,000 feet. It now proceeded towards Spalding, a 17 minute flight. The 306th continued to follow its leader, high and on the right.[34]

Major Normand and Colonel Peaslee had no idea where the other was located, but by now they both had broken radio silence in order to place each other. Neither leader was successful in raising the other on his Very High Frequency (VHF) radio.

The 305th was now on the right of the 92nd and 306th. They were separated by about 14 miles, visibility was still restricted, and they were still too far away to see and identify each other.

Normand maintained his heading to the northwest, and at 1148 his group crossed the briefed flight route that Peaslee was flying to Spalding. Normand and his 17 B-17s from the 305th flew two more minutes and were now ahead of Peaslee. At 1150 he turned right to a heading of 52 degrees and headed for Spalding.[35] The three groups of the 40th Combat Wing were now converging on this third checkpoint. The 305th was at 12,200 feet and still climbing.[36]

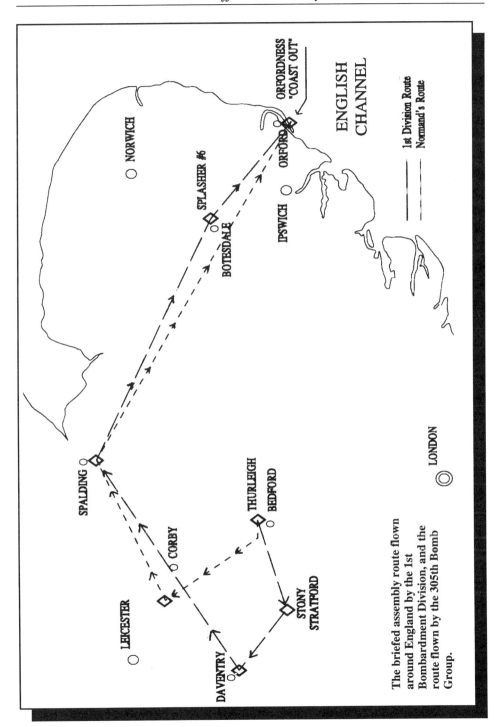

The briefed assembly route flown around England by the 1st Bombardment Division, and the route flown by the 305th Bomb Group.

As it turned out, the 305th cut across well in front of the 92nds route to Spalding at about a 90 degree angle. The 305th was now on the left and far ahead of Peaslee's two groups.

At 1153 hours, the 92nd arrived at the intersection where the 305th had crossed in front of it just five minutes earlier. The 305th, still ahead and on the left, had not seen any other aircraft since leaving Thurleigh. On the Holt ship, the word was passed to the crew, "Be alert and report if you spot any of the wings!"[37]

At 1159 Normand passed over Spalding seven minutes *early* at an altitude of 15,000 feet – 1000 feet below his assigned altitude. To run time off the clock, he had Lieutenant Kane execute a 360 degree turn to the left which normally took two minutes.[38] Normand again tried his VHF, but could not get through to Peaslee, the 40th Combat Wing leader. However, he did see a complete, three group, combat wing formation in the area.[39] Since the wing he saw was not missing a group, he correctly assumed it was not his, the 40th. However, apparently he had forgotten his wing was leading the 1st Division today and everyone else would be waiting for and then forming on Peaslee when he arrived at the various checkpoints. The wing he now saw off in the distance was the 41st Combat Wing awaiting the arrival of Colonel Peaslee. Normand still had no idea Peaslee and the 40th Combat Wing were still minutes behind him. If Major Normand had waited over Spalding for his and the 40th's briefed departure time at 1206, it would have allowed the 40th Combat Wing to draw nearer, and the two would have finally met. As it turned out, Peaslee, still looking for the absent 305th, was four miles south of Spalding at 1205.[40] Peaslee reached that checkpoint on time, at 1206, at an altitude of 15,000 feet, which was the same altitude Normand was flying.[41]

However, the get-together was not to be, and the 305th leader missed a golden opportunity to be at the right place at the right time. At 1201, *five minutes early*, Normand apparently became impatient and ordered Kane to depart Spalding. The 305th now picked up a southeasterly heading of 130 degrees directly for Orfordness, the last possible briefed gathering point. This heading would take him just south of the next to last check point called Splasher #6. Normand continued to maintain 145 mph IAS.[42]

Upon arrival at Spalding, the two groups of the 40th Combat Wing made a right turn to 130 degrees. They, too, were now heading for Orfordness via Splasher #6 but five minutes behind the 305th. Peaslee continued to maintain 150 mph IAS.[43]

At 1223, Peaslee and Normand reached the vicinity of Splasher #6. Both were at 17,000 feet but five miles apart. Peaslee had now passed Normand and was several minutes ahead of him.[44] However, due to layers of clouds and rain in the area visibility was quite restricted and neither one saw the other.

Lieutenant Edwards, Normand's navigator, stated, "... We attempted to call the 92nd which was leading but received no answer, although we could hear them at varied intervals. As we approached Splasher #6 we heard the leader say he was going below the overcast and I therefore assumed he was still actually below the overcast and we would be able to join them when they broke through ...We left the English Coast at Orfordness, 2 minutes late and without having joined our Wing ..."[45]

The 1st and 41st Wings were now in trail behind Colonel Peaslee's 40th CBW, and the 1st Division rendezvous was complete. To the south, the 3rd Division was gathering, while to the north the B-24s of the 2nd Division were also trying to form. However, due to adverse weather conditions, the B-24s now planned a diversionary mission north along the Dutch coast.

The 1st Combat Wing, which was following behind Peaslee, had a problem identical to his. Lieutenant Colonel Milton, its leader, was still missing his low group, the 381st, which had inadvertently met Peaslee's two groups over Thurleigh earlier. Fortunately for the 1st Combat Wing, its leader and the 381st were in radio contact and making plans to meet later on. Thus, while the 381st was racing to catch up to its wing, the 305th leader pondered his problem of being all alone with no radio contact with his wing leader.

Many of the crew members in the 305th aircraft were now beginning to wonder what was going on. They were all alone, and none of them had observed any other aircraft since they were assembling over Thurleigh.

In the low squadron, Sergeant Arthur E. Linrud, McDarby's engineer, figured the mission was not going according to plan and they would probably be recalled to base.[46] Sergeant Bill Frierson and several others on the Holt crew were surprised to be alone and were wondering if the mission had been aborted.[47] Lieutenant Ken Kenyon, the low squadron leader, was optimistic, believing the 40th CBW was ahead of them. He thought surely the 305th would soon catch up.[48]

The high squadron's leader, Lieutenant Farrell, was puzzled as to why he had not seen any other groups for a long time and Lieutenant Fred Helmick, Farrell's bombardier, was extremely concerned the outfit was all alone.[49] Staff Sergeant Jayson Smart, at the right waist position, had it figured correctly, "They were all there, behind us!"[50]

In the lead squadron where the decisions were being made by Normand, things were just as bad. On the Bullock crew, Technical Sergeant Carl Brunswick was hoping the mission would be scrubbed.[51] Staff Sergeant Joe Kocher, on his 23rd mission, was getting a very lonesome feeling wondering what was happening. He was used to seeing all the other formations after assembly, and he asked himself, "Are we going out there all alone?"[52] In Normand's aircraft, morale was not any better. Technical Sergeant LeRoy V. Pikelis, the engineer and top turret gunner, figured they were lost and the 305th was late getting to the bomber stream. After listening to a number of conversations on the intercom between Joe Kane, the pilot, Jack Edwards, the navigator, and the group leader, Normand, Sergeant Pikelis came to the conclusion Major Normand was having a bad day – for making decisions.[53]

The three squadrons maintained radio silence between themselves, so no information was being passed around. Major Normand continued to try to contact Colonel Peaslee on VHF but to no avail.

At 1231 hours the 40th CBW was two miles north of Orfordness at 18,000 feet and about a minute late. Here it made a 360 degree turn to allow the 1st and 41st Combat Wings a bit more time to close up the 1st Division formation.[54]

At 1232, flying at 18,500 feet, the 305th approached Orfordness. While the 40th was executing the 360 degree turn to the north, the 305th passed to the south of Peaslee by some four to five miles and neither saw the other. So near and yet so far! Both now headed out over the English Channel toward the enemy coast, and the 305th was now well ahead of everyone else!

Today's estimated time enroute to cross the channel from Orfordness to the Dutch coast was 22 minutes. However, as it turned out, it would take Normand and the 305th Bomb Group 32 minutes to make the trip.[55]

Chapter 5

Across The Channel

The briefed route for the 1st Division from the English coast to the European continent was to be flown on an east southeasterly heading of 108 degrees. This would enable the bombers to cross the Dutch coast in between Walcheren and Schouwen, the southern most islands of the Zeeland area. The distance was 81 nautical miles, and the estimated time en route was 22 minutes.[1]

Major Normand, still well ahead of the 1st Division, now ordered Lieutenant Kane to proceed on course as he took the 305th, all by itself and in good formation at 18,500 feet, out over the channel and headed for the Dutch shoreline.

Shortly after departing the English coast, the 305th Group and Major Normand suffered another setback. Lieutenant R.E. Reid, flying in the number five position of Normand's lead squadron, did a 180 degree turn and headed back towards England. The crew had been on oxygen since climbing through 10,000 feet and now Lieutenant Reid alerted his crew members those in the cockpit were not getting any air to breathe. Staff Sergeant William E. Lutz, the ball turret gunner, had already left his turret complaining about breathing problems.[2] When it became apparent the oxygen system was malfunctioning, Lieutenant Reid returned to Chelveston, but he could not land due to poor visibility. At 1412 hours he was able to land safely at Great Ash Airfield.[3]

Now down to 16 aircraft and all by itself, the 305th continued across the English Channel still trying to find the rest of the 1st Division bombers. Normand kept looking to his front for Peaslee, and he continued to come up empty.

Normand was still unaware that Colonel Peaslee trailed him by several minutes and figured the 40th Combat Wing together with the rest of the 1st Division was somewhere out in front of him. He also told Kane to increase his climbing airspeed from 145 to 155 miles per hour in order to catch Peaslee.[4]

When Colonel Peaslee arrived at Orfordness at 1231 to lead the division across the channel to the target, he still was hoping the 305th would be waiting for him at the "coast out." Such was not the case. Peaslee had no idea where his missing low group was located, so he decided to go on without it.

As the bombers of the 1st and 3rd Divisions broke free of the English coast, the solid undercast began to dissipate, and the poor visibility that had caused so many earlier problems began to rapidly improve. In just a matter of minutes, both units were now flying in clear, sunny weather. The 2nd Division now began its flight to the north towards the Frisian Islands.

Today the waters of the channel appeared calmer than usual. The lack of white caps on the frigid, dark green waves indicated a mild sea with little wind. As the 1st Division continued across the channel, some crew members throughout the various groups took a long look at the sea below them. A number of them had flashbacks of how the chilling waters of the channel had left them with traumatic experiences they would rather forget. As the lone 305th Bomb Group continued its slow climb, several airmen aboard its aircraft also remembered the harrowing times they had spent in that cold water while trying to survive.

Flying low squadron lead today, Lieutenant Kenyon recalled how he and his crew had been extremely fortunate on 9 September 1943. While returning from a raid into France, they had run low on fuel and were forced to ditch their aircraft in the middle of the English Channel into a choppy sea with extremely large swells. As it hit the water, the aircraft struck a wave and broke in half at the ball turret. The ship sank rapidly, but everyone was able to scramble out quickly into the two, five-men life rafts. The crew managed to tie the two dinghies together so they would not drift apart. They floated aimlessly for an hour and a half before being picked up. Fortunately for them, the British Air Sea Rescue boat arrived minutes before a German vessel which had been racing out from the enemy coast to capture them. British Spitfire fighters also arrived on the scene, and they proceeded to completely ruin the Germans' day. The rescuers were so far out it took them two hours to feel their way back through the British mine fields.[5]

Fortunately for Technical Sergeant Richard W. Lewis, Kenyon's regular waist gunner, he was not aboard that day. However, as Lewis gazed down at the water from the right waist window, he remembered he had already been dunked three days earlier, on 6 September, while filling in with another crew. The outfit he flew with was returning from Stuttgart, Germany, where they had received considerable flak damage. They made it to within five miles of the English coast when, due to fuel exhaustion, the pilot alerted the crew for ditching. With the exception of the pilot and copilot, the remainder of the crew took its routine escape positions in the radio compartment. When the aircraft hit the water, Dick Lewis took a hard blow to the head and face. The dinghies inflated and the rest of the uninjured crew began scrambling out the radio hatch into them. By this time Lewis was covered with blood, unconscious, and left for dead. British Air Sea Rescue arrived almost immediately, and began taking aboard the downed flyers. The B-17 remained afloat despite the tail

Members of a bomber crew who successfully ditched in the English Channel about to be picked up by the British Air Sea Rescue Service. (Air Force magazine photo, January, 1945)

being torn off on impact. While rescue operations continued, Lewis regained consciousness, inflated his "Mae West",[6] and managed to pull himself through the open waist window into the water. He was immediately fished from the channel by the British – wet, cold, and with a bellyful of salt water. With the empty fuel cells of the aircraft providing a buoyancy, the aircraft was still floating as the crew left the scene in the safety of the British boat. As they headed for home, the tired band of Americans watched from a distance as British Spitfires were forced to sink the B-17 with machine gun fire.[7]

Over on the right, in the high squadron, Lieutenant Clinton A. Bush was flying copilot today for the Fisher crew. He, too, now relived an experience where he had felt the touch of the cold, deadly water which now lay far below him. On his third mission in July, 1943, he was substituting as copilot with a different crew. While over the target, at Saint Nazaire, France, the aircraft took numerous hits from antiaircraft ground fire. The ship caught one burst of flak underneath the left wing which punched a large hole in a fuel cell – too large for the tank to seal itself. As the pilot headed for home, raw gasoline could be seen leaking from the wing in a fine spray. Over the channel a small wing fire started. In just moments, the entire left wing was ablaze. The pilot opted not to have anyone bail out and land in the near freezing water which normally provided a survival time of only ten to twenty minutes. Instead he alerted British Air Sea Rescue, and made a procedural, crosswind landing approach to the water. Both British fighters and a surface craft were waiting for the B-17 when it splashed down in a trough between waves. It was a good landing, and when the aircraft settled into the water everyone began to exit. The pilot slipped out his left side window while Bush pulled himself through the one on the copilot's side. The remainder of the crew began clambering out of the radio compartment hatch on the top of the fuselage. As they gathered on the wings, there was momentary confusion. Both life rafts had released as designed; i.e., one from each upper side of the fuselage adjacent to each wing. However, the left dinghy was badly burned from the wing fire, and the right one was full of flak holes and would not inflate. In addition, burning gasoline floating on the surface of the water began to surround the disabled aircraft. To make matters even worse, the tail gunner had gone into shock, refusing to leave the radio room. The aircraft now began to rapidly fill with water, and there was no way to reach the British boat. Acting quickly, members of the crew pulled the emotionally disturbed gunner out of the radio hatch, and then the entire crew inflated their life jackets and jumped into the cold water. They managed to push their way through the burning fuel, but they were too far away to swim to the British high-speed launch. The crew members then all locked arms, formed a circle, and waited. This prevented them from being individually swept away by the swift current, and it also provided the British with the best possible rescue scenario. The rescue boat eased in alongside the wet and cold flyers and quickly pulled them to safety. The tail gunner's buddies kept him afloat until all were rescued.[8]

At 1244, twelve minutes after the 1st Division bombers started departing the English coast, the Luftwaffe began to stir. Their fighter pilots had already been alerted for almost an hour. They first put up six single-engine squadrons, including three from the south Holland-Ruhr area that were flying west to meet the bombers.[9]

At 1247 another German radio message to its fighters was intercepted by friendly forces. It indicated the Allied force was between Orfordness and the Dutch islands at an estimated altitude of between 20,000 to 23,000 feet.[10]

After the 1st Division departed the English coast and while well out over the English Channel, Colonel Peaslee arrived at a strange and gutsy solution to the problem that had been plaguing him since departing Thurleigh – the whereabouts of the missing 305th Bomb Group. Rather than lead the 1st Division with only two groups, Peaslee decided he would tie in his 40th Wing behind another combat wing and proceed in the center of the division formation.[11]

In his book, *Heritage of Valor*, Colonel Budd Peaslee wrote, "There are orders in Eighth Air Force to bomber commanders to abort the mission rather than attempt to penetrate the German defenses with less than a complete wing formation. The 40th is short one-third of this force [the 305th]."[12]

Thus his decision: strange in that he was improvisiong a type formation had not yet been employed in combat, and gutsy in that he was in direct violation of an apparent Eighth Air Force directive. So rather than abort and take the 92nd and 306th Groups home, he elected to stay and fight.

Believing the 41st Wing was directly behind him Peaslee called its leader, Lieutenant Colonel Rohr, to take over the point. However, the 1st Combat Wing was trailing Peaslee, where it was supposed to be, with the 41st behind it. There was momentary confusion for the 1st Division. Colonel Peaslee then contacted Lieutenant Colonel Milton, the 1st Combat Wing commander, assuming Milton had a complete wing of three groups.[13]

"I'm short one group," Peaslee said. "You will take over the lead. I'll fly high on you, and I'll retain air command from that position."[14]

Milton acknowledged the radio transmission. However, he now had second thoughts about who should be in command, and he felt he had an immediate decision to make. Since he (Milton) was now authorized to lead shouldn't he also take over the responsibility of command? He decided he should and went ahead and assumed command of the 1st Division from there on.[15] It is not known whether Milton told Colonel Peaslee he was usurping his authority and taking command, but Peaslee later denied ever giving Milton command of the 1st Division.[16] Milton's presumptuous attitude in "assuming command" was indeed difficult to justify. The 1st Division now had two air commanders!

Also at this time, Lieutenant Colonel Milton and his 1st Combat Wing were short a group, the 381st. Apparently Milton felt the order not to penetrate the enemy coast with only two groups did not apply to himself. While Milton, with only two groups, the 91st and 351st, moved forward to secure the lead, Peaslee began to maneuver his two groups, the 92nd and 306th, in behind Milton. The 41st CBW, with the 379th Group leading, the 303rd high on the right, and the 384th low on the left, continued to follow the other two wings.

Milton's 1st Combat Wing was well south of Peaslee and considerably behind him when the lead began to change hands. In order to get behind Milton's formations, Colonel Peaslee had to actually backtrack his two units by flying some ten miles to the southwest to position himself.

At 1251 hours and just 25 miles short of the enemy coast, Peaslee began an ill-fated maneuver. At this time, his navigator's log indicates the 92nd started a large "S" turn so as to fall in behind the 1st Combat Wing.[17] At 1255, after four minutes of maneuvering, Colonel Peaslee finally moved in behind and on the left of the 1st CBW. These turns, over the channel, were unusual and this poorly planned exercise disrupted the formations of Peaslee's two groups and caused them to spread out and become ragged. The move to swap places with the other wing may have been a good one, but it did not work out that way for Peaslee and his 40th Combat Wing.

Meanwhile, the 3rd Division continued to follow, south of and slowly converging on the 1st Division, but the B-24s from the 2nd Division were nowhere to be seen. Due to the weather, and unknown to Colonel Peaslee, the 2nd Bombardment Division would not participate in today's mission.

It was evident the 305th leader, Major Normand, either never heard of the above order referred to by Peaslee, or his inexperience allowed him to ignore it. He continued to stay in front of the 1st Division and lead his group of 16 B-17s, all by themselves, across the English Channel directly toward the enemy's shoreline.

Even though he could not see any B-17s in front of him, Normand remained on course. The 305th was now only 16 miles from the enemy coast as it continued its climb and now passed through 19,000 feet. Sergeant Pikelis, Normand's top turret gunner, had the best range of visibility in the lead squadron, and he had been constantly traversing his turret in 360 degree circles hoping to see someone friendly out there besides the 305th. This time as he swung the turret laterally from the nose towards the tail he saw elements of the 1st Division behind, high, and to his right. They were all climbing. The improving visibility left no doubt they were B-17s and lots of them. He quickly broke the silence of the intercom, and told his leader where the rest of the bombers were located.[18]

At approximately the same time Peaslee swung in behind Milton's 1st Combat Wing, 1255 hours, Normand was eight miles ahead and three miles south of Milton. Normand now decided he had better join the others. He had his pilot, Joe Kane, execute a turn to the left to a heading of 65 degrees to intercept the 1st Division,[19] which was on a 120 degree course. He also had Kane decrease his indicated airspeed from 155 to 145 mph.[20] This deceleration now caused a chain reaction for some of the 305th pilots following Normand. A number of them had to decrease power very rapidly in order to slow down and avoid overrunning those in front of them! When that problem was overcome, the pilots then had to react quickly by adding power to their heavily loaded aircraft so as not to approach stalling airspeed, and at the same time avoid falling too far behind the formation.

As visibility continued to improve, Normand searched for his wing that was missing one group. The leading wing of the bomber stream was now only five miles away and on his left. He could see clearly now, and this first outfit only had two groups in its box. He assumed it had to be the 40th Combat Wing which he had been chasing all day – only it was not! Unknown to the 305th leader, the 1st CBW had taken the lead, and the 40th had moved in behind it and was now running second.

The entire 1st Division was now flying in clear weather and had unlimited visibility. This was further evidenced by the observations of several 305th crew members. Shortly after Normand started his initial turn to cross in front of the oncoming bomber stream, up in the high squadron, Lieutenant Bush, Fisher's copilot, leaned forward and looked back to his left and out the pilot's side window. He observed B-17s of the 1st Division, which were still approaching the 305th from the northwest, spread out as far as he could see.[21]

As the turn continued, down in the low squadron, Sergeant Lester J. Levy, right waist gunner on the Murdock crew, finished test firing his weapon. He walked over to the left waist position which was now facing southwest in the turn, and leaned over the shoulder of Staff Sergeant Tony E. Dienes. Together they took a long, admiring look at the White Cliffs of Dover, which were clearly discernible from over 65 miles away.[22]

The copilot of the Bullock crew, Lieutenant Homer L. Hocker, watched as Normand's aircraft, just off his right wing, continued in the left turn. Hocker now looked left and could see the other wings approaching the 305th, and he figured his group would keep circling and wait for the bomber stream to reach them. He, like so many others in the 305th, had not planned on playing "catch up" after the completion of this turn they were now executing.[23]

The 305th Group now cut across the path of the oncoming 1st Division. Since he was still well ahead of the bomber stream, Normand had Lieutenant Kane start another turn to the left, planning to roll out of it and join the lead wing as it came by. Normand continued in the turn while his other 15 ships tried to stay in position and maintain a close group formation. He misjudged his timing and the rate of closure of the oncoming 1st Combat Wing, and it proved to be too much for the 305th. By the time the group had completed a very awkward 360 degree turn, which took much too long, the 1st CBW had gone roaring by.[24] The 305th was now behind and far out on the left side of Colonel Milton's wing. During this gyration, both the high and low squadron formations of the 305th began to loosen up, and string out.[25]

Normand, who had decelerated to 145 mph from 155 mph during this maneuvering over the channel, now tried to play "catch up" by increasing airspeed to 150 mph.[26] Many of the heavily laden aircraft in his high and low squadrons, still struggling to recover from the unorthodox turns, could not keep up with these constant airspeed changes, and they began to fall further behind.

Major Normand pressed on, determined to overtake "his" wing! As a result, the 305th formation was now becoming badly dispersed and disorganized. The low squadron was well behind the rest of the group. The high squadron, having been on the outside of the 360 degree turn, also continued to fall behind. Lieutenant Robert A. Skerry, flying #4 position in the high squadron, kept dropping back, could not close it up, and was one of the first ships to fall behind.[27] Ahead of him and on the right, Flight Officer Fisher's aircraft, also in the high squadron, could not keep up and steadily began dropping behind his leader. Staff Sergeant Loren Fink, Fisher's left waist gunner, could see the lead squadron far ahead of his ship and the only other B-17s he saw were all off to his right.[28] The Fisher and Skerry aircraft and many other 1st Division B-17s would not be in formation when the enemy arrived. Lieutenant Charles W. Willis, Jr., on Skerry's right, stayed with him as they formed a two-ship formation of their own.

At 1259, just five minutes from the Dutch coast, Major Normand, using

The returning friendly P-47 fighter convoy from Eighth Fighter Command reported many of the bombers were considerably strung out while being escorted along the route. (Photo courtesy of Clarence R. Hall)

excessive power settings, was still trying to catch up to the two groups that had passed him while the 305th was executing the turns.[29] Also at this time, Lieutenant Edwards, the 305th navigator, recorded in his log the arrival of friendly fighter support.

The 353rd Fighter Group appeared with 44 P-47s, each equipped with 75 gallon belly tanks for extra range. Despite the extra tanks, the fighters had already consumed valuable fuel, as the bombers were ten minutes late completing the rendezvous.[30]

The fighters had their hands full trying to cover the entire 1st Division, as the "little friends" reported the bomber boxes (formations) were considerably strung out.[31] The earlier bad weather over England, coupled with the just completed wing leader changes and Normand's maneuvering, had taken its toll on the close, effective formation flying expected from the bomber pilots.

The lateness of the bombers, plus the ragged formations, caused many in the B-17s not to see the fighters – even though they were there. Lieutenant Colonel Milton, the new leader, did not see the P-47s at all, but thought he saw contrails way up high.[32] The 92nd Group was five miles inside the Dutch coast when Colonel Peaslee's navigator recorded seeing the fighters.[33] The leader of the 41st Combat Wing, Lieutenant Colonel Rohr, saw contrails of the P-47s about fifteen miles off the enemy coast and 5000 feet above the bombers.[34] In the 305th Group, Lieutenant Gerald B. Eakle, flying in #5 position of Kenyon's low squadron, took a quick glance upward as four, four ship sections of P-47s slid across at a 45 degree angle from left to right in front and well above the squadron.[35] Some crew members in other 305th aircraft never saw any friendly fighters.

At 1300 hours, at least 12 German twin-engine aircraft took off from the Baltic coast area, and an additional two squadrons of them were reported flying south from the Quackenbrück, northern Germany area.[36] The Germans continued to systematically launch aircraft to counter the forthcoming American attack!

At 1301 hours, flying at 20,500 feet, Major Normand finally caught up with "his" wing.[37] As he slid into the low group position, much to his dismay he realized it was not the 40th CBW, but the 1st Combat Wing consisting of the 91st Bomb Group led by Lieutenant Colonel Milton with the 351st flying high group on the right. Normand slowed to Milton's speed, elected to stay put, and took over the low group position which belonged to Milton's absent 381st Bomb Group.

Normand had now arrived at the wrong place at the wrong time!

Just minutes later the "missing" 381st Group pulled up on the left of the 1st Combat Wing to take its position as low group in the formation. However, it was now occupied by Major Normand and the 305th. More confusion! Making the best of a bad situation, the leader of the 381st, Major George Shackley, took a high position on the right of his "sister" outfit, the 351st.[38] Instead of three groups, Milton now had a four group wing.

As the history of the 1st CBW reads, "... Fortunately for us, the 305th Group, which was unable to find its own gang, came along with us and flew low box on our incomplete wing formation. This they did at their own expense..."[39]

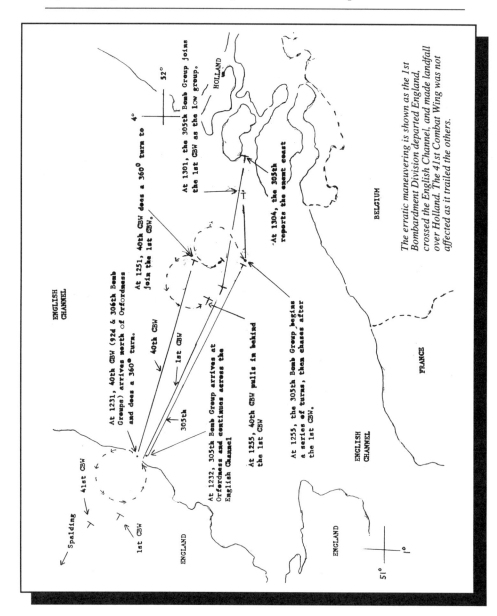

The erratic maneuvering is shown as the 1st Bombardment Division departed England, crossed the English Channel, and made landfall over Holland. The 41st Combat Wing was not affected as it trailed the others.

Chapter 6

Into Holland

When the 353rd Fighter Group, consisting of the 350th, 351st, and 352nd fighter squadrons, approached Walcheren and Schouwen Islands, high above, and on the flanks of the 1st Division, a gaggle of 20+ Bf 109 single-engine German fighters was sighted flying at 32,000 feet. They approached the P-47s from 11 o'clock and began attacking the American fighters.[1]

The tactics of the enemy aircraft were slightly different than in the past. They seemed to concentrate solely on the fighters and disregard the B-17s entirely. It appeared the strategy was to break up the fighter convoy so as to leave the bombers unescorted further along the route.

As the Germans attacked, the 352nd Fighter Squadron, guarding the division's left flank, turned to engage and circled in behind the enemy. Half the Bf 109s peeled off and dove for the deck, while the others stayed to fight. The P-47s destroyed four of the attackers and damaged three. In the meantime, the 351st fighters split into sections. Half of them moved to assist the 352nd, while the reminder covered the rear box, the 41st Wing. The remaining 350th fighters moved forward to cover the 1st and 40th Combat Wings.[2]

At 1304 inaccurate flak was encountered from the coastal batteries at Domburg and West Schouwen by the 1st Combat Wing as it crossed over the Dutch coast at 21,000 feet. The American force was now entering Holland just north of course over Schouwen Island and many of the 1st Division planes were still struggling to catch up and get in formation.

The 1st Combat Box (wing) now consisted of the 91st Bomb Group in the lead, the 351st high on the right, the 381st high on the right of and above the 351st, and elements of the scattered 305th flying low on the left. Discounting the 305th, the 1st CBW had initially launched 49 B-17s, but 16 of them had aborted for various reasons and now this lead box was down to 33 ships.[3]

The 40th Combat Wing, which was following the 1st, still consisted of the 92nd and 306th Bomb Groups. Between the two of them, out of 39 planes that took off, they had five abortions, so this second box now contained 34 B-17s.[4] Again not including the 305th, both wings contained almost the same number of aircraft, but there was one big difference. From a defensive posture, the 1st had a tight box, while the 40th did not.

The 41st Combat Wing continued to bring up the rear with 48 ships in good formation. This unit had nine aborted aircraft from the original 57 that took off.[5]

At 1305 hours the Germans launched five more squadrons of single-engine fighters from the Laon area located northeast of Paris. They flew west anticipating a deep penetration.[6]

The 353rd Fighter Group was continuously engaged by many groups of enemy fighters as the 1st Division continued through the Dutch airspace. So far, the enemy had not bothered the many B-17s of the 1st Division that had fallen behind and were not in formation. Especially fortunate was the 305th which was well scattered out over Holland. At 1309 Lieutenant Joseph E. Chely trailing behind and below Major Normand in the #4 position in the lead squadron of the 305th Group, began to lose power on #1 engine. He had experienced the engine "running rough" while assembling over England, but it had cleared up somewhat and he had continued on. At this time, the group was in a steady climb, and his airspeed fell off gradually from loss of power. He began to slip further and further behind Normand's lead element. Lieutenant Victor C. Maxwell, flying on Chely's left wing, stayed with him as long as he could, but he had to eventually leave Chely or he would have never caught back up to the lead squadron.[7]

Chely's aircraft now began to vibrate, as it continued to lose power. His engineer, Technical Sergeant J.D. Balsley, advised his pilot the engine had internal damage that could not be rectified from the cockpit. Lieutenant Chely made one more attempt to clear the engine but had no success. The 305th leader, unaware of Chely's predicament, continued to pull away from him. By now the 305th Group was well into Holland and still disorganized.

At 1315 hours Chely, realizing he could not continue the mission, broke radio silence, called Normand and told him he had to abort. Since a number of other lagging B-17 aircraft were near him and not guarding his radio frequency, Chely lowered his landing gear to inform them he was leaving and heading for home.[8]

After clearing himself, he broke out of the aggregation and did a 180 degree turn back towards England. Despite being chased for a time by a lone German fighter, the Chely crew returned safely to Chelveston. The Chely aircraft touched down at 1429 hours. Inspection of the number one engine revealed a broken exhaust stack on #7 cylinder.[9]

With Lieutenant Chely's departure, the 305th now had three aircraft abort, and the group was reduced to 15 ships. At the same time Chely aborted, 1315 hours,

Normand's navigator entered a position report in his log. Major Normand was 19 miles ahead of Chely when the lieutenant turned around for home.[10]

Lieutenant Maxwell, now back with the lead squadron, had been in the #6 position all day. With Chely and Reid gone, he was now all alone so he took over the #4 position vacated by Chely, so the normal three ship, second element of Normand's lead squadron now consisted of one B-17 flown by Lieutenant Maxwell. Behind Normand the high and low squadrons were not in a close group formation.

So far Major Normand had experienced a bad day. In the morning he failed to identify the groups circling over the wing assembly area at Thurleigh. Then he did not follow his briefed routes, arrival and departure times, and airspeeds to the other checkpoints over England.[11] At 1232 hours, he decided to lead his lone group from the English coast across the channel to enemy territory. Shortly thereafter, he attempted several prolonged, unorthodox turns over the channel in order to join the wing following him which did not work out as planned. Finally he elected to chase this wing (the 1st Combat Wing) which he thought was the 40th.[12] This series of blunders by Major Normand, while over the channel, spread out many of the group's aircraft and severely disrupted the defensive posture of the 305th which was shortly to meet the enemy.

The 1st Bombardment Division continued on course. Milton adjusted his bearing further to the southeast, to a heading of 122 degrees, as the division passed across southern Holland and just inside the northern Belgian border in the area of Antwerpen. The antiaircraft batteries positioned around the city of Antwerp fired for several minutes at the B-17s, but the flak was inaccurate and caused no damage. They were inside Belgium for about 15 minutes.

At 1324 a German formation arrived in the border area where the bombers would penetrate Germany, and the leader called his ground control, "Shall I wait here because of Allied fighters or shall I fly on?"[13]

He was told to wait for the Americans.

At 1328 the division was about to reenter Holland where that country's southern corridor joins with the borders of Belgium and Germany. Also at this time several more squadrons of enemy fighters arrived in the sector between Malmedy, Belgium, and Koblenz, Germany, and were ordered to wait in that area.[14]

German fighter control also encountered a problem at 1328. The 3rd Bombardment Division, which had been on a converging course from the south to meet with the 1st Division in the Aachen area, had now been detected. The enemy suddenly realized the Americans were conducting a double incursion. The Germans had to quickly decide how best to employ this force now positioned in the Malmedy-

Koblenz region. Finally at 1342 a determination was made to use it against the 1st Division![15]

At about 1330 hours 30 Fw 190s made their first attempt to attack some of the bombers, but they were unsuccessful. They employed tactics of splitting formation, so half could attack the bombers from head-on while the other fighters came up from below. At the same time 20 more enemy fighters were seen coming down toward the bombers from 34,000 feet. Both enemy forces were driven off by the escorting P-47s.[16] The "little friends" were busy and effective in protecting the "heavies" from the Germans, but in just a matter of minutes the convoy of fighters would have to depart for lack of fuel.

Also at 1330, more Allied radio intercepts indicated another 100-120 enemy fighters, from the Enschede-Leeuwarden area in northern Holland, were just 23 miles east of the 1st Division and closing fast. An additional three enemy squadrons were being brought south from Oldenburg, near Kiel, while three more from the Deelen sector, in south central Holland just northwest of Arnhem, were also on their way.[17]

From the amount of German fighter control radio traffic being picked up by the Allies, it was a foregone conclusion the enemy was mustering every operational fighter it could get into the air. This was later reconfirmed when Allied radio monitors began picking up exchanges between German ground stations and five night fighter units, thereby adding another 100 more enemy fighters for the Americans bombers to take on.

It was now 1332 and many of the P-47s began to rock their wings in a final salute to the B-17s as they began turning around for home. They could do no more for them.

Suddenly the "little friends" were gone, and the B-17s were all alone in the skies over Europe.

Weather-wise, it was an ideal day over the continent. Warm and mild, sunny, a few scattered cumulus clouds, and unlimited visibility. In fact, it would have been a beautiful day to go flying – if there hadn't been a war on!

The German fighters continued to gather for the assault. Lieutenant John A. Cole, the 305th navigator for Kenyon in the low squadron, could see many fighters coming up to meet the American planes.[18] Over in the 305th high squadron Lieutenant Stanley Alukonis, Lang's copilot, looked up ahead and saw great numbers of fighters massing to attack – anywhere from 50-100. To him they looked like a swarm of bees.[19] Also in the high squadron, Lieutenant Helmick, Farrell's bombardier counted 40 German fighters silhouetted against stray cumulus clouds climbing to engage them.[20] Before the day was over, 325 single-engine and 165+

Into Holland

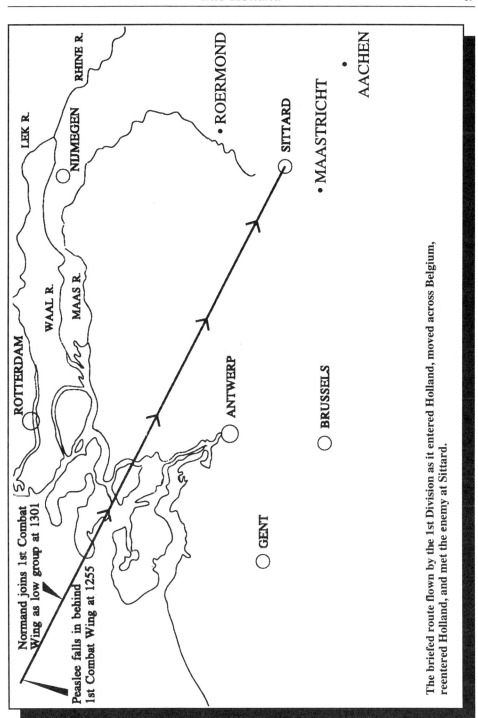

The briefed route flown by the 1st Division as it entered Holland, moved across Belgium, reentered Holland, and met the enemy at Sittard.

twin-engine German fighters would attack the 1st and 3rd divisions.[21]

Sittard, a town in the south of Holland, lies just over a mile west of the German border, and four miles east of the Belgian border in the narrow corridor that separates these three countries. It was here the Germans began their first attacks on the now unescorted bombers of the 1st Division, and the fighting would continue uninterrupted for the next three hours and fourteen minutes.[22]

At 1333 hours the German fighters made their move, and the 305th group lead navigator, Lieutenant Jack Edwards, recorded in his log, "fighter attack!"[23] Edwards also noted a major shift in heading. For some unexplained reason, Lieutenant Colonel Milton now changed from his briefed southeasterly course of 122 degrees and headed northeast on a bearing of 72 degrees.[24] Normand followed. At this time, the 305th's four ship lead squadron still clung to its position low, behind, and on the left of the 1st Combat Wing. However, the 305th's high and low squadrons had still not managed to get into their proper group formation defensive posture as the fighting began. They were by themselves still trailing Normand. To further compound the problem, each of these squadrons had several individual aircraft out of formation, and these unfortunate crews had become stragglers.

The German pilots, as always, were looking to take advantage of any weak links in the bomber boxes. The first attacks, mainly by single-engine aircraft, were concentrated on the front box containing the three groups of the 1st Combat Wing and stragglers of the disconcerted 305th Group.[25] As more German fighters arrived in the battle zone, especially the twin-engine fighters, they proceeded to the second box and began to clobber Colonel Peaslee's scattered 40th Combat Wing.[26]

The 41st Combat Wing which continued to follow Peaslee, was not immediately touched by the battle. Lieutenant Colonel Rohr, who was leading the 41st with his 379th Bomb Group, watched the battle continue in front of him for almost thirty minutes before his outfit received their first attacks.[27]

Lieutenant Colonel Milton's 1st Combat Wing maintained a close defensive formation, and was spared much of the initial brunt of the onslaught against the first box. The spread out 305th Group was most vulnerable and was given top priority by the enemy.

Single-engine Fw 190s and Bf 109s began their assaults almost simultaneously on Kenyon's low squadron and Farrell's high squadron. They came at both squadrons head-on, singly, in pairs, and using four and six ship elements. As soon as one element completed its run past the American ships, it was immediately followed by another.

Many large groups of fighters came in waves. Helmick, in the high squadron's lead ship, later described some of the action, "After they [40 German fighters] hit us head-on, another group of about 40 hit us head-on again!"[28]

Some individual fighters passed through the formations, broke right or left, climbed back up to altitude and prepared for another charge. Many that made frontal attacks would do a half roll just in front of the bombers and dive away while others executed a barrel roll right through the formations.[29] Lieutenant John C. Tew, Jr., Lang's navigator, watched as one Fw 190 roared through the high squadron so close he could see the face of the pilot and the white scarf he wore – not more than 20 feet away.[30]

As different German units arrived on the scene, tactics varied somewhat. Putting the sun at their backs, some single-engine fighters attacked from the rear, six o'clock high position while others came from the right and left rear.[31]

One group of fighters broke up into five and six ship sections and flew on the left and parallel to the 305th just outside the gunners range. Still paralleling the bombers, the fighters would move out well ahead of them and then come around in a gradual right turn, rolling out on a head-on collision course flying directly at them. One flight section would follow the other as each swept through the scattered B-17 formations.[32]

Ten minutes after the single-engine fighters began working over the 1st Division bombers, many of the twin-engine German fighters arrived to join the fight. Over seven different models of these aircraft were seen, predominantly Ju 88s and Bf 110s.

Their maneuvers were different from the single-engine aircraft. While the single-engine fighters attacked mostly from the front and sides, the twin-engine aircraft would stand off 1500 yards to the rear and lob rockets and heavy cannon projectiles into the formations. On several occasions, these rocket firing aircraft flew to within 200 yards of the rear of the formations before launching their missiles. Some of them moved in even closer as they fired their 20mm cannon. Several times they made frontal attacks from the ten to two o'clock level positions. They came singly and in elements of three, four, and five. Their rockets were installed under each wing and when fired emitted a long, red streak followed by a burst similar to flak.[33]

The 1st Division B-17 crews did not lack for targets. There were so many of the enemy around them, many of the gunners on the crews stopped calling out the fighter positions on the ship's intercom. They just picked targets and fired away.

The Allies continued to monitor the air battle via radio. At 1339, six minutes after the battle began, they heard many enemy fighters reporting to their ground control they had only 15 to 20 minutes of flying time remaining, and they began requesting landing instructions for refueling.[34] As soon as they were cleared to leave the area, other aircraft reported on the scene and took their places in the fight.

As they arrived on station, some were ordered to go for the first box, while others were directed to the second wing led by Colonel Peaslee. He, too, like so many other commanders, was having trouble keeping his formations together, while the Germans were doing everything they could to disperse any existing close defenses of the American units.

Ammunition was beginning to present a problem for the bomber crews. Each plane had taken off with approximately 6800 rounds on board, but the gunners were firing so often at so many targets they were expending their .50 caliber machine gun ammunition at an alarming rate. Crewmen began cautioning each other to pick a good target, let the enemy get in closer, and use short bursts. In the 305th, Staff Sergeant Jayson C. Smart, Farrell's right waist gunner, kept telling the crew to save ammo as it was still a long way to the target and home![35]

As if this was not enough for the crews to handle, at 1350 hours German ground control passed the word to its fighters who had not yet joined the clash and thought the American P-47 fighters might be in their district, "Heavy bombers east of Aachen, without fighter escort!"[36] This was the signal for those fighters to make straight for the raiders.

At 1400 hours a follow up message was transmitted to verify the P-47 situation. German ground control now told its fighters, "Allied fighters have turned back, heavy bombers are alone!"[37]

The Germans were now free to concentrate all available aircraft against the B-17s of the 1st Division!

Chapter 7

At The Border

The first several flights of attacking single-engine German fighters caused considerable damage to many of the straggling B-17s in both the high and low squadrons of the 305th Group, with the low squadron initially taking the brunt of the pounding. It wasn't long before the enemy aircraft began picking off the American bombers, one by one, in the second elements of each of these squadrons.

When the attacks began at 1333 hours, the high squadron was at 22,000 feet, and the low one was at 21,400 feet. Far ahead of these two units was Major Normand's four ship lead squadron at 21,700 feet following Lieutenant Colonel Milton's 91st Group. Any semblance to a 305th group formation was nonexistent – it was now every squadron for itself!

The Murdock Crew

Lieutenant Douglas L. Murdock's B-17, flying #6 in the Purple Heart corner of the 305th low squadron, was hit hard in the first pass by several Fw 190s which made head-on attacks firing their 20mm cannons.

There was a terrific explosion between #1 and #2 engines, and the cowling was blown off #1. Both Murdock and Lieutenant Edwin L. Smith, the copilot, were stunned, and they momentarily blacked out from the force of the blast. Smith, who was flying the plane at the time, was the first to recover.[1] He realized the ship had lost boost on both of these engines as the superchargers were inoperative, causing a great reduction in power. The aircraft could not keep up with the low squadron and began to fall further behind. Lieutenant Smith could not maintain altitude. He started a diving turn to the right to get down on the deck and away from the constant fighter attacks. As the aircraft descended through 20,000 feet, Lieutenant Murdock recovered from his temporary shock and told Smith he would take over flying the plane.[2]

As the enemy kept up the pressure, Murdock headed south and continued to lose altitude. He was now much lower than the other bombers and came under attack from three Fw 190s led by *Major* Seifert, of Jagdgeschwader 26 (USAAF equivalent to the *Jagdgeschwader* was a Fighter Group).[3] They came in from the tail, and Staff Sergeant William B. Menzies claimed a hit on one of them as they went by.[4]

Sergeant John W. Lloyd realized he was not getting any oxygen in his ball turret position, had to evacuate it, and came up into the waist section. Staff Sergeant Tony E. Dienes, left waist gunner, helped Lloyd out of the turret and began administering oxygen to him.[5] As Dienes returned to his left waist position, he was immediately hit by fragments of a 20mm cannon shell and fell to the floor wounded.[6]

More cannon fire hit the bomber, causing it to roll and pitch downward, tossing Sergeant Lester J. Levy, the other waist gunner, past the ball turret and all the way forward to the radio compartment wall. His parachute, which had been lying next to his feet, was also thrown up against the wall. He got to his feet, grabbed his chute, and went aft to aid Dienes. In doing so, he accidentally unplugged his intercom system just as Ted Smith, the copilot, ordered the crew to bail out![7]

Those in the front of the aircraft were also busy. Murdock continued to fly the aircraft and lose altitude and seemed dazed and disoriented. The aircraft was being shot to pieces, so Smith ordered the crew out.[8]

While kneeling over the unconscious Dienes, Levy looked up to see the radio operator, Technical Sergeant Thelma B. Wiggens, Jr., and Lloyd moving to the waist door position putting on their chest packs to bail out.

Levy, still unaware the crew had been ordered to bail out, asked what was going on and Lloyd yelled at him, "Get the hell out of here"![9]

He moved quickly, found his chute, and scrambled to the waist door. In his haste to don it, he only hooked the left clamp of the parachute to the left ring of the harness he was wearing. He did not realize the tacking that held the right ring in place had ripped loose and was hanging down from the harness. He was the first to reach the waist door, but in his hurry to leave he forgot to pull the door's emergency quick release handle. Instead, Levy kept trying to force the door open and forward against the slipstream and was having little success. While he was trying to keep the door from blowing shut, the loose, right ring on his harness flew up and hit him in the eye. From an equipment standpoint, he was suddenly reminded he was not completely ready to jump. He let the door slam shut as he quickly snapped the loose, right side together.[10] In the meantime, someone behind him reached forward and pulled the handle releasing the waist door – and out the three of them went!

As some of the other crewmen began to leave the aircraft, Murdock directed Staff Sergeant John E. Miller, the togglier, to salvo the bombs. However, Sergeant Miller apparently never heard the pilot's order, for the bombs remained aboard.

Murdock continued flying on a southerly heading still losing altitude and trying to evade more fighters. As Smith headed down into the navigator-bombardier compartment to bail out via the nose hatch, he glanced back at his pilot, and Murdock appeared to be okay and getting ready to leave the ship.[11] Smith also noted that Technical Sergeant Russell J. Kiggens, the engineer, had just stepped down from the top turret seemingly alright.[12]

When Smith left through the nose escape hatch, Lieutenant John C. Manahan, the navigator, and Miller, the togglier, had already bailed out. Smith assumed Kiggens was right behind him and would follow him out.

Murdock's bomber passed over the northern outskirts of Maastrich, Holland still chased by Major Seifert. The German pilot broke off the engagement as the B-17 rolled over on its left wing and began to spin.

At 1340 it crashed near Limmel, Holland, a small village just north of Maastricht and several miles east of the Belgian border.[13] Shortly after the crash, the bomb load exploded.[14]

The McDarby Crew

Lieutenant Dennis J. McDarby had been playing "catch up" since the mid-morning group assembly over England. Once again he was out of his #3 position behind the low squadron with the low squadron still trailing Major Normand. McDarby was still struggling to recover from Normand's earlier, rapid acceleration that had left the high and low squadrons behind and far too spread out.

He later reported, "Before I was shot down, we never did catch up to any wing and join with them. We were all by ourselves."[15]

The McDarby crew had now become a straggler and many German fighters swarmed to this lone B-17. As the top turret gunner, Sergeant Arthur E. Linrud, turned in his turret he could see that no part of the formation was escaping attack.[16]

His aircraft began shaking violently after the initial head-on German attacks. The #1 engine took a cannon hit. It lost oil pressure so rapidly the propeller could not be feathered, setting up a vibration throughout the ship. McDarby's left window was shattered from the enemy fire, but fortunately he only suffered superficial wounds.[17]

The second wave of fighters also came in from 12 o'clock level pouring more cannon fire into the bomber. Engine #2 was hit, exploded, and also could not be feathered.[18] Due to the windmilling, uncontrollable propellers, these "runaway" engines caused even more shuddering as a third attack came from both sides.

One of the attackers, 23 year old *Feldwebel* (Sergeant) Helmut F. Brinkmann, flying a Bf 109 from the 7./JG 1 (USAAF equivalent, 7th Wing, 1st FG), got in too close and was hit by fire from several of McDarby's gunners.[19] Brinkmann's aircraft immediately went into a steep dive. It completely disintegrated as it slammed into the ground near the village of Haanrade, Holland just inside the Dutch-German border. The pilot's remains were never found.[20]

The Germans began another assault. This time a 20mm cannon shell exploded above and behind Staff Sergeant Domonic C. Lepore in the tail compartment, wounding him in the head.[21] Left waist gunner Sergeant Leonard R. Henlin, on his fifth mission, collapsed at his post from more exploding enemy cannon fire. He was last seen sitting dead at his waist position.[22]

A small fire now started between #1 and #2 engines, spread quickly, and engulfed the entire left wing. McDarby called the crew and ordered them to bail out, and they wasted no time in leaving. All of them were fully aware raw gasoline in the fuel cells of a burning wing could explode at any moment.

Lepore left by the tail exit, while Staff Sergeants Hosea F. Crawford, the radio operator, and Benjamin F. Roberts, ball turret gunner, moved to the waist door. They were immediately followed by Sergeant Robert G. Wells, the right waist gunner. Crawford and Roberts managed to get out, but Wells was killed from another fighter attack before he could jump.[23]

Up on the flight deck, McDarby told Sergeant Arthur Linrud, the engineer, to go down into the nose and open up the nose hatch. Linrud went below, released the door, and as he started to climb back up onto the flight deck, Lieutenant McDarby reached down from his seat, tapped him on the head, and told him to bail out.[24] Sergeant Linrud went back down into the nose and prepared to exit through the hatch. He saw both the navigator, Lieutenant William J. Martin, and Lieutenant Harvey A. Manley, the bombardier, getting ready to follow him.[25]

The pilot began to leave the cockpit and saw his copilot, Lieutenant Donald P. Breeden, crawling into the bomb bay to bail out.[26] The aircraft was at an altitude of 15,000 feet when McDarby went down into the nose of the bomber and bailed out.

As the B-17 fell the left wing broke off, and at approximately 1345 hours the aircraft came down in two pieces near the small, Dutch coal mining village of Eygelshoven, Holland, which is on the border with Germany.[27]

The Eakle Crew

The Eakle B-17 was in #5 position of the low squadron when the attacks began. The initial German charge was from both sides by three ship elements of Fw 190s but caused no damage.²⁸ Then came the head-on assaults as numerous flights of fighters kept coming through the low squadron formation. A 20mm projectile hit the leading edge of the left wing and passed completely through the airfoil without exploding. However, it penetrated a fuel cell, and raw gasoline was now trailing from the wing. A string of 20mm shells then exploded on the left between #2 engine and the fuselage causing extensive damage to the plane. An oil line to #1 engine was severed causing a "runaway prop" condition, and #2 engine began to smoke.²⁹

The left side of the nose had been shattered, and Staff Sergeant Herman E. Molen, the togglier, was cut on the neck by a piece of flying Plexiglas. Lieutenant Charles B. Jones, the navigator, received wounds in both legs from the fragments of the same exploding cannon round.³⁰

The Germans kept coming, and the bomber continued to accept battle damage and fly. Three fighters coming out of the sun made a run at the left side of the ship. Staff Sergeant Alfredo A. Spadafora, at left waist, claimed one of them that was going down in flames. He now glanced up just as the Dienhart aircraft passed overhead trailing raw gasoline from a large hole in its right wing.³¹

An intercom check between the pilot, Lieutenant Gerald B. Eakle and Staff Sergeant Lloyd F. Knapp, Jr., right waist gunner, indicated the waist area of Eakle's airplane looked "like a sieve"! The Germans kept pounding away, and the B-17 still stayed in the air. Back in the tail section the gunner, Staff Sergeant Lloyd G. Wilson, reported the tip of the right horizontal stabilizer had been shot off. He could also see a large hole in the left wing just outside the engine.³² Gerry Eakle and his crew mates were now all by themselves unable to keep up with the remainder of the squadron. Eakle now did a diving 180 degree turn and headed for home, but he would never make it!³³

The next attack by the German fighters knocked out the oxygen and electrical system, with #3 engine taking a hit that blew off the cowling. That engine began to smoke then caught fire.³⁴

Spadafora could see the engines smoking, large holes in the wings, and now he realized there was no intercom. He recalled, "There was no longer a formation – our plane was all torn up and going down!"³⁵

Technical Sergeant Barden G. Smith, the engineer, came out of the top turret and into the cockpit to assist his pilot in getting some oxygen. As he handed the "walk around" bottle to Eakle, a round of 20mm exploded in the upper turret destroying it.³⁶ The somewhat shaken Smith had just stepped down into the right place at the right time! It was time to abandon ship and since the intercom was out Eakle rang the alarm bell as the plane descended through 10,000 feet.

The tail gunner, Sergeant Wilson, heard both waist gunners talking, then his intercom went dead. He realized the oxygen system was out, and the electrical system was apparently inoperative for his heated suit was not working. He noted five Fw 190s trailing his ship. Three of them had red painted cowlings while the other two had yellow noses with brown stripes. However, for some reason, they were not attacking.[37] Molen, up in the nose, also noticed the attacks had stopped and figured the enemy was just waiting for them to bail out.[38]

Wilson now observed four parachutes open behind the plane, and three of the Fw 190s made passes at the men hanging in the chutes. Since Wilson had not heard the order to bail out he tried to contact the pilot on the intercom, but it was still inoperative. Wilson now left the tail section and went forward. Both waist gunners, Knapp and Spadafora, plus Technical Sergeant Donald H. Norris, the radio operator, and the ball turret gunner, Staff Sergeant Robert L. Sanchez, were all gone.[39]

As the tail gunner passed through the waist section and radio compartment, he was amazed at the battle damage he saw – there were holes and damage everywhere. He continued forward then entered the bomb bay, and watched as Smith, the engineer, pulled the emergency release and salvoed the bombs in order to lighten the ship.[40]

At this time Molen pushed up through the doors dividing the nose compartment from the flight deck, and the copilot, Lieutenant Walter H. Boggs, told him to bail out. The togglier ducked back down into the nose section, got the wounded navigator, and helped him up onto the flight deck. Herman Molen then pulled the conscious but bleeding Jones through the "upper local" gun position and onto the catwalk of the bomb bay. There he gave Jones' parachute a quick, last check and then pushed him out of the plane through the open bomb bay doors. The aircraft was about 1500 feet above the ground as Molen immediately followed Lieutenant Jones out the opening.[41]

Lloyd Wilson, still standing at the rear end of the bomb bay catwalk, had watched Jones and Molen leave, and he now looked on as the engineer, Barden Smith, followed them. Tail gunner Wilson was not sure exactly what he wanted to do, but he was certain he did not want to jump![42]

He now moved forward to the flight deck, and saw the pilot, Eakle, and copilot, Boggs, getting ready to bail out. Wilson asked the pilot if he was going to crash-land the aircraft. Eakle, still trying to stay airborne so the crew could get out, told the bewildered gunner to go ahead and jump![43]

Wilson went back into the bomb bay and straddled the opening. He continued to wait. He still did not want to jump. Once again he made his way back to the flight deck, where he saw Eakle heading for the nose escape hatch, and motioning for Wilson to jump. The copilot, Boggs, was getting up out of his seat when the plane lurched violently and threw him back into it.[44]

The tail gunner now followed the pilot down into the nose compartment. Wilson still waited! Again Eakle motioned for Wilson to get out, and the gunner finally bailed out at approximately 1000 feet. He was immediately followed by his pilot.[45] The burning aircraft continued on, and crashed moments later at approximately 1345 hours just across from the Dutch border near Eisden, Belgium.[46]

The Willis Crew

Lieutenant Robert A. Skerry, who had moved into the #4 slot of the high squadron during the assembly over England, had been one of the first to fall behind his unit ever since Major Normand had decided to chase after the 1st Combat Wing.[47]

Skerry had been desperately trying to close the distance but still had not succeeded. Lieutenant Charles W. Willis, Jr., Skerry's right wingman stayed with him during this ordeal, and they were all by themselves as a two ship formation when the fighting began.[48]

Fw 190s and Bf 109s came at them singly and in pairs from all directions. The fighters attacked head-on, from both sides, and from the rear.[49] The single-engine fighters were later joined by several twin-engine Bf 110s that fired rockets at the two bombers from the side, nine o'clock position.[50]

Willis' B-17 was hit immediately and fell off in a diving turn to the left as Skerry continued ahead chased by numerous fighters. Willis was pursued by several single-engine fighters which pounded his plane relentlessly. Despite control damage and loss of power, he managed to roll out of the turn but continued to maintain a rapid descent. The fighters kept on attacking, further disabling the aircraft.

Willis continued to lose altitude rapidly, and the pilot now ordered everyone to bail out, and for Lieutenant Willard E. Dixon, the bombardier, to salvo the bomb load. This Dixon did and then followed the navigator, Lieutenant Christian W. Cramer, through the nose hatch and out the ship.[51]

Staff Sergeant Alan B. Citron, the radio operator, entered the waist section just as Staff Sergeant Bernard E. Snow came up out of his ball turret. Both donned their chutes, headed for the waist exit, and jumped.[52]

Immediately after Citron and Snow left, the right waist gunner, Staff Sergeant

John L. Gudiatis, was severely wounded from an exploding 20mm shell, and he collapsed on the floor of the ship.[53] He had been hit on the right side of his body, receiving wounds in his leg, thigh, buttock, and upper portion of his arm and shoulder. He was bleeding heavily as he snapped on his chute and crawled toward the waist hatch. The wounded flyer noticed the left waist gunner, Staff Sergeant Edward J. Sedinger, whose back was towards Gudiatis, was still standing at the waist exit and had made no effort to jump.[54] As John Gudiatis continued to pull himself along the floor to the exit, he saw Staff Sergeant Nicholas Stanchak, the tail gunner, lying in the tunnel leading from the tail position to the waist. He could see Stanchak's face was bloodied, and he was unconscious. Gudiatis began to drag himself on his stomach over to Stanchak to help him and another round went off near him, shattering his right hand. As he reached Stanchak, he passed out.[55]

The left waist gunner, Sergeant Sedinger, still had not jumped. He had refused the order to bail out and continued to remain in the waist section. The engineer, Technical Sergeant Floyd J. Karns, went back through the bomb bay and into the waist to confront Sedinger.[56]

Karns tried to order Edward Sedinger out, but the waist gunner would not jump. They argued and Karnes realized he could not force him out, so he left and went forward to the cockpit.[57]

As Sergeant Karns entered the flight deck, he saw Lieutenant Willis still in the pilot's seat trying to control the aircraft so the crew could jump. Lieutenant John J. Emperor, the copilot, motioned Karns to go out the bomb bay, then he got out of his seat and reached for his chest pack. The engineer entered the bomb bay and prepared to jump. He turned and saw the copilot standing on the forward end of the catwalk clutching his "popped" parachute. Somehow the chute had opened accidentally as Emperor was crawling through the top turret to the bomb bay. Karns could wait no longer and jumped.[58]

Willis had been flying on a northeasterly heading as he came down. He now made a 90 degree turn to the northwest, passed directly over the Maas river, and overflew the towns of Horn, then Baexem. A six-year-old Dutch boy, Leo Zeuren, in Baexem watched as the Willis aircraft now did a 180 degree turn and headed southeast back towards Horn. The German fighters continued to attack the bomber as it rapidly lost altitude, passed back over Horn and began to recross the Maas river. It now appeared to witnesses on the ground Lieutenant Willis, while still under attack by six to eight enemy fighters, was attempting to make a crash landing between the village of Horn and city of Roermond.[59]

Willis' aircraft was about 1000 feet above the river when people on the ground

A B-17 staggers away from its formation with an engine on fire. (Air Force magazine photo, February, 1944)

observed six crew members parachute from the plane. Willis was now very low as he throttled back and began making an approach to an open field. The fighters still continued to harass him even at tree top level. As he neared the ground, his right wing clipped the tops of several large trees. The aircraft veered to the right, crashed and burned at approximately 1345 hours.[60] The crash site was in Holland between the towns of Horn and Roermond and some four miles west of the German border.[61]

Chapter 8

Into Germany

The fighting back and forth across the Dutch border ended in just twelve minutes for the 305th Bomb Group with the loss of four aircraft. The battle now moved into Germany.

The enemy continued to concentrate on the dispersed 305th and the spread out two groups, the 92nd and 306th, of Colonel Peaslee's 40th Combat Wing. It was now impossible for the 305th to regroup.

The elements of the leading 1st Combat Wing, also under attack, were continuing to maintain their close, tight formations. This would pay them tremendous dividends as the unrelenting, running conflict continued throughout the day.

Just prior to the outbreak of hostilities, the 1st Division leader had been maintaining a southeasterly heading of about 122 degrees.[1] As the fighting began at 1333, Milton, leading with his 1st Combat Wing, made a left turn off course to the northeast to a bearing of 72 degrees.[2] Beginning with this heading and for approximately the next 20 minutes (1333 to 1353), Lieutenant Colonel Milton would lead some elements of the 1st Division off course to the northeast then alter his course back to the southeast. During this time frame, Milton would skirt the edge of Cologne, then execute a sharp turn to the south and bring the division (minus) almost directly over the city of Bonn.[3] At 1353, a final left turn south of Bonn would bring Milton back on the correct southeasterly course.[4]

At 1348 one formation of B-17s, Milton's 1st Combat Wing, was reported coming over Cologne by the enemy's fighter control.[5] Colonel Peaslee, trailing some miles behind Milton with his 40th Combat Wing, had not followed him when he strayed off course but continued to maintain the prescribed southeasterly heading.[6] Lieutenant Colonel Rohr, bringing up the rear with the 41st Combat Wing some 10 miles behind Peaslee, had also stayed on course. He noticed that up ahead Milton's groups seemed far to the left of course and were receiving quite a bit of flak.[7]

It is not known whether or not the intensity of the enemy fighter attacks caused this change in course. Whatever Milton's reasons were for this deviation from the flight plan, it only added more problems to be concerned with for many of the bomber crews of the 1st Division. As Milton led the bombers adjacent to Cologne and then

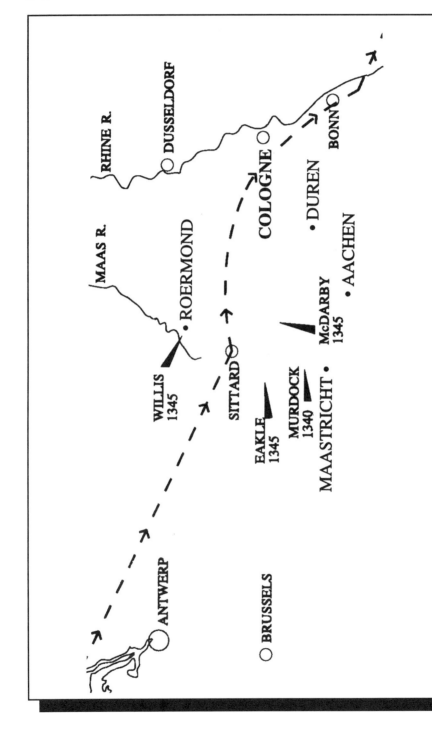

turned south and headed for Bonn, he brought them within range of numerous flak batteries deployed around the two cities![8]

The German antiaircraft units, which had been tracking the B-17s for some minutes, were waiting patiently for the Americans to come within range of their weapons. The bombers soon obliged, and the Germans began filling the sky with 88mm and 105mm rounds.

Several crews of the lead squadron of the 305th, which were still on the left of Milton's 1st Combat Wing, began receiving flak which was moderate to intense and very accurate. Normand's ship received some minor battle damage, and Sergeant Pikelis, up in the top turret, saw two B-17s at two o'clock high (Milton's outfit) get rocked by close flak.[9] Lieutenant Kincaid, flying on Normand's right wing, received numerous holes throughout the aircraft, and the right wing tip disappeared.[10] Lieutenant Bullock's aircraft, over on Normand's left wing, somehow remained untouched.[11] Also at 1348 much of the fighting now shifted to the area between Aachen and Cologne.[12] Many of the German aircraft now moved to engage the 40th and 41st Wings which had stayed on course. Quite a few of the fighters were making their final attacks prior to breaking off their initial engagements with the bombers, as they needed to land in order to refuel and rearm.

The low squadron of the 305th Group was now down to three ships. Lieutenant Kenyon, the leader, still had with him Lieutenant Holt off Kenyon's right wing in the #2 position, and Lieutenant Dienhart trying to stay in the #4 slot behind and below Kenyon. Despite the constant attacks since the fighting started, these three crews still managed to fight back, maintain a close formation, and remain airborne.

As it approached Cologne, what remained of the this low squadron continued to follow Major Normand's lead element, but Kenyon and his other two ships were still well behind it. Kenyon, Holt, and Dienhart now began to feel the heat from the flak batteries around the edge of that large city. Both the Kenyon and Holt ships caught several close bursts, but they suffered little if any damage.[13] Dienhart's ship staggered from a number of damaging hits. One burst of 88mm exploded under the right wing, followed by a blast just in front of the nose.[14] To make matters even worse, the German fighters, ignoring the bursting antiaircraft fire, continued to keep the pressure on the three ships. The Bf 109s and Fw 190s continued their head-on attacks, and twin-engine night fighters now joined in lobbing rockets at the planes from the rear of the formation.

Lieutenant Colonel Milton leading with the 1st Combat Wing, with Normand still in the low group's position, continued south and passed over Bonn. In just a few minutes he would finally get back on his briefed route.

The Holt Crew

Every gunner on Lieutenant Robert W. Holt's crew was up and firing. Despite the ship taking numerous hits, everyone aboard was intact.

Tail gunner, Staff Sergeant Mike S. Letanosky, was scoring big. He had just shot down one fighter, and now he got another. The fighters had attacked from six o'clock low coming up in a slight climb. Both Bf 109s were confirmed by the ball turret gunner, Sergeant William C. Frierson, who saw them start smoking and spin away! He continued to watch as they fell, and he saw one of them explode. There was no chute from the downed fighter, but Frierson, who could see far below, saw the sky was filled with parachutes from both sides.[15]

Staff Sergeant Frank W. Rollow firing from the left waist position, also started an Fw 190 smoking as it passed by.[16] The engineer, Staff Sergeant Charles E. Blackwell, manning the top turret, observed a fighter take hits and parts begin to fly off – then it, too, began to smoke.[17]

The navigator, Lieutenant Edward O. Ball, came on the intercom with, "Come on gang, let's get them", followed by, "that one got us!"[18]

Suddenly the entire situation changed for this crew as two more fighters came in from the 12 o'clock level position. Cannon fire raked the nose section and cockpit. Immediately following were another pair of Bf 109s whose 20mm rounds literally wiped out the front portion of the aircraft killing the pilot, Lieutenant Holt, and severely wounding his copilot, Lieutenant Aubrey C. Young, Jr.[19] Down in the nose section, the navigator, Lieutenant Ball along with the bombardier, Bryce Barrett, also died from the same fighter attack.[20]

As the last two fighters completed their run and closed on the bomber, they both did a half roll sliding underneath, down, and away. Frierson, in the ball turret, picked up one of them as it broke underneath the bomber. He scored several hits, and the Bf 109 began to smoke.[21]

Immediately afterwards, from the corner of his eye, the gunner also caught sight of a door and a body pass by his position. The left wing of the B-17 dropped, and the ship started a shallow turn. He now swung his turret from nine o'clock to three o'clock to check the engines – everything looked fine. The B-17 continued in an aimless, left turn and began to lose altitude.

Frierson decided to check upstairs. He swung his turret to the six o'clock position, rolled the guns straight down, and quickly came out of the turret hatch. He immediately saw Staff Sergeant Floyd A. Lenning, the right waist gunner, and Rollow, the other waist gunner.

Both were standing by the waist escape hatch watching Letanosky, and the radio operator, Technical Sergeant Doris O. Bowman, trying to kick open the jammed waist exit.[22]

The ball gunner went into the radio room, picked up his chest pack, and snapped it in place. The radio room door to the bomb bay was open, and he glanced towards the cockpit. He could clearly see the copilot, Lieutenant Aubrey C. Young, Jr. lying face down across the top turret pedestal.[23]

Immediately after this last attack, the intercom system went out, and Blackwell, the engineer, was wounded in the right leg by fragments from one of the 20mm shells that had ripped into the cockpit.[24] He remained in the top turret standing on the steel pedestal and firing his twin .50 calibers until he felt someone trying to crawl through his legs. He came down out of the turret and stepped over the lifeless form of Lieutenant Young. Apparently the badly wounded copilot had died trying to get out of the cockpit into the bomb bay.[25]

Sergeant Blackwell also found a raging oxygen fire consuming everything in sight. There was so much flame he could not even see the pilot. He realized he needed to find his chute and get out of there. The last he remembered he had placed his parachute on the floor behind the pilot's seat. The heat was so intense he had to cover his face and just grope and feel for the chute. Fortunately, he found it immediately. He then quickly slid down through the aperture separating the two pilots' seats and entered the nose section. It was also completely ablaze, and Blackwell again had to shield his eyes and face from the heat. He could not see anywhere and could not make certain if either the navigator, Lieutenant Ball, or Lieutenant Bryce Barrett, the bombardier were lying hurt or had already bailed out. The heat and fire were becoming unbearable. He managed to feel his way along the left side of the nose section, found the escape hatch, jettisoned the door, and pulled himself out through the opening.[26] The engineer and the nose hatch door were what ball turret gunner Frierson had seen go by his station just moments before.

Sergeant Bill Frierson returned to the waist section from the radio compartment and approached the others who were still trying to open the waist door. Bowman called over to Sergeant Lenning and asked him to help them open the exit.[27]

All of them were unaware the cockpit had been wiped out, and there was no one up front to give the bail out signal. They did realize it was a "no win situation", and it was time to abandon ship! The waist exit would still not budge.

The left wing of the pilotless aircraft dropped at a much steeper angle, and the ship began a slow spin. The five gunners in the waist were now pinned down by centrifugal force and unable to move. The rotation of the aircraft increased as it continued to spin and lose altitude. As hard as they tried, none of those trapped in the waist were able to overcome the force holding them down, and they could not crawl or pull themselves to another exit.[28] It was just a matter of moments until the bomber would slam into the ground.

Abruptly, as the plane neared the ground, the tail section snapped. It partially broke loose, then finally separated completely from the aircraft just aft of the waist section.[29] The aircraft was approximately 800 feet above the terrain and was now falling in two large pieces with a gaping hole at the end of each segment.[30]

In the tail section, Frierson, Rollow, and Lenning were immediately thrown on top of Sergeants Bowman and Letanosky pinning them down at the bottom of the pile.[31] There was not much time left, and everyone was struggling to clear the falling wreckage. Sergeant Rollow was on top of the pile, and he scrambled to get out the new found exit, but his feet were hung up with those of Lenning. Frierson, underneath both of them, somehow managed to untangle the two releasing Rollow who rolled free and out the opening. He was immediately followed out the hole by Lenning. Frierson's feet became caught in some control wires, and he could not get them untangled. He quickly loosened his prized pair of British flying boots, freed his feet, and pulled his rip cord. His chute opened, and he was free of the wreckage – and his footwear.[32]

At 1350 the front portion of the burning aircraft crashed just outside the village of Immendorf, Germany, approximately 11 miles northeast of Aachen.[33] Moments later the tail section came to rest several hundred yards from the forward portion of the ship. Shortly after the crash the bomb load exploded.[34]

The Dienhart Crew

Just like the others in the low squadron, Lieutenant Edward W. Dienhart's aircraft had been taking a terrific pounding from the German fighters since the air battle began at 1333 hours. Earlier during the fighting he had continued to stay in formation, in the #4 position, tucked just behind and below the leader at an altitude of 21,300 feet.[35]

Dienhart and his crew had all they could handle. Things continued to go from bad to worse as the Germans pressed home their attacks vigorously. Exploding 20mm cannon fire from head-on fighter assaults had shot out the manual throttle control cables regulating the engine speeds for #1 and #2 engines.[36] Dienhart now had no way of adjusting engine power on the port side. However, the B-17 power plants had a built-in safety valve for such an occurrence. They automatically went to full power – so all was not lost. The #3 inboard engine on the copilot's right side was still undamaged and running smoothly, but #4 began to loose oil pressure, had to be shut down, and the propeller feathered.[37]

During the early fighting, Dienhart had trouble controlling his ship, and it had drifted left. He corrected the left drift, and then it slid back to the right and had passed over the top of Eakle's aircraft, which had been flying on Dienhart's right wing. Sergeant Spadafora, Eakle's left waist gunner, could see raw gasoline draining from a large hole in the right wing of Dienhart's aircraft.[38] The Eakle aircraft, also under heavy attack at the time, had then fallen away from the formation.

As Dienhart's ship passed near Cologne, flak exploded in front of the aircraft and took out a good portion of the Plexiglas nose of the B-17, severely wounding the bombardier, Lieutenant Carl A. Johnson, and Lieutenant Donald T. Rowley, the navigator.[39] The intercom had also been shot out, so the pilot had no way of knowing what had taken place in the nose section.

Most of the fighter attacks were from head-on at the 12 o'clock position, but the tail gunner, Staff Sergeant Bernard Segal, was extremely busy at his six o'clock position. The first few attacks on his station came from five and seven o'clock level, and just outside the maximum left and right deflection of his twin .50 caliber machine guns.[40] However, the German pilots grew bolder and became careless. On their third pass, four Fw 190s came right at Segal from the level position. As they closed the gap to about 100 yards, the tail gunner began to score as one of them completely disintegrated, and a second began smoking furiously and fell away.[41] As the other two fighters whipped by, Bernie Segal looked right and left of his position and realized both his pilot's wingmen, Murdock and Eakle, were gone. Up front, Dienhart noticed Holt's aircraft began to lose altitude and then began a gradual, descending left turn. Staff Sergeant Robert Cinibulk, Dienhart's right waist gunner, reported to the cockpit they were the only other ship left in the squadron besides the leader, Kenyon![42]

Staff Sergeant Christy Zullo firing from the left waist observed a German pilot bail out of his smoking fighter, while down in the ball turret, Staff Sergeant Raymond C. Baus saw two other German pilots leave their ships.[43] Enemy planes were everywhere. There was so much exchange of gunfire it was difficult, at times, to ascertain which gunner was scoring against which fighter.

More blasts shook the plane from exploding flak. This time two 88mm rounds ripped through the radio compartment. One buried itself in the floor and came to a stop in the camera well, but it did not go off.[44] The other detonated wounding the radio operator, Technical Sergeant Hurley D. Smith, in the right side and both arms.[45] Since the intercom was not working, Smith conscious but hurting had no way to notify the crew he had been hit.

Just seconds later, a 20mm cannon round went off in the waist section between the two gunners. Both Zullo on the left and Cinibulk on the right felt the fragments hit their flak suits. Fortunately, neither one of them was injured.[46] It was amazing they had not been wounded throughout the fighting, for the waist section was riddled with holes from one end to the other. To make matters worse, the oxygen system on the left side of the ship now failed.

Down in the ball turret, Sergeant Baus, was having his share of problems. His turret had been hit, rupturing the hydraulic line, and squirting oil all over him. The turret slowed, then stopped completely – it was jammed. He tried to rotate the door hatch of it up into the plane, but the turret was fouled and would not operate. He then took the hand crank, called for help from above, and tried to manually rotate the turret to the exit position. Every time he cranked the turret it slipped a bit off its rotating mechanism. Baus began to wonder if he and the turret were going to separate from the aircraft. Above him both waist gunners were also working furiously to help him get the door hatch properly positioned. Finally the hatch was in place and out crawled the shaken, oil soaked gunner.[47]

Up in the top turret, Technical Sergeant George H. Blalock, Jr., the engineer, was having trouble with his two .50 caliber machine guns. One quit firing completely, while the other would only work sporadically.[48]

Dienhart, still unaware he had three badly wounded crewmen aboard, decided his crew and bomber had taken enough punishment. He tried the intercom one more time to order the crew to abandon ship. Again it did not work. He now flipped the toggle switch to ring the alarm bell so everyone would bail out. The pilot next motioned to the copilot, Lieutenant Brunson W. Bolin, and Blalock, the engineer, to bail out.[49] They left their positions, put on their parachutes, and went down into the nose section to leave. Bolin saw the bombardier and navigator lying there wounded, and he started back up to the cockpit to tell Dienhart of their condition. However, he never got a chance. As Lieutenant Bolin started back up onto the flight deck, Dienhart waved frantically at him to bail out. Bolin obeyed the order and therefore was unable to tell his pilot there were two wounded men in the nose. He and Blalock then left through the nose exit.[50]

Lieutenant Dienhart continued to fly for another minute or two to enable the rest of the crew to clear the plane. By some act of God, Dienhart now decided *he* would *not* bail out but would stay with the ship.

At approximately 1349, several minutes after Holt fell out of formation, Dienhart rolled the B-17 into a sharp left turn from 21,000 feet and headed for the deck, still *unaware* he had those three badly wounded crewmen aboard. This left the Kenyon crew, the last ship of the low squadron, all alone as it struggled to survive the overwhelming fighter attacks. As Dienhart put the bomber into the steep dive, his tail gunner, Segal, looked up from his position and could see Kenyon's aircraft was just ahead of his plane and in trouble.[51]

Chapter 9

Approaching the Rhine

The four-ship high squadron of the 305th was still trying to catch up to Normand's lead squadron. Lieutenant Farrell, its leader, continued pulling excessive engine power, but it did not help much to close the gap.[1] Lang was off Farrell's left wing but having trouble staying with him in formation.[2] Flight Officer Verl D. Fisher was trailing behind and out of the number #2 position off on Farrell's right. Staff Sergeant Loren M. Fink, Fisher's left waist gunner, could see the lead group of the 305th far ahead of them.[3] Lieutenant Skerry was still airborne and well behind Fisher. For all of them, somewhere up ahead lay the Rhine river.

The Fisher Crew

Like so many other stragglers singled out for attack, it was now the Fisher crew's turn to get burned. Fw 190s and Bf 109s came at Fisher's B-17 head-on, abreast in six and eight ship sections.[4] Fisher lost power on one engine and began to have trouble controlling the aircraft. Staff Sergeant Harold Insdorf, down in the ball turret, took a direct hit from a cannon shell, and his position suddenly became very quiet.[5] The radio operator, Technical Sergeant Harvey Bennett, called Insdorf on the intercom and received no reply. Bennett immediately left the radio compartment and went aft to the ball turret. He called again to the ball turret gunner, and there was no answer. He began to pound on the turret, when he noticed a gaping, six inch hole at the top of it. Harold Insdorf had been hit in the head and died inside the turret.[6] The fighters continued to rake the bomber with cannon fire. Sergeant Fink, at the left waist position, was hit and knocked to the floor with a shattered right shoulder. At the same time, Bennett, still trying to see if Insdorf might be alive in the turret, was wounded in the left arm and leg.[7]

The pilot continued to have extreme difficulty controlling the aircraft. The intercom was now inoperative so Fisher rang the alarm bell for the crew to abandon ship and told Lieutenant Clinton A. Bush, the copilot, to get the bomb bay doors open and pass the word to bail out.[8] Lieutenant Bush began to get out of his right seat and in the hurried process of reaching for and putting on his chest pack, he accidentally snagged the chute's D-ring. This was a large, steel forged, cadmium plated "D" shaped handle used to activate the chute. As he moved out of his seat, Bush did just that. He inadvertently "popped" his parachute all over the cockpit.[9] He now managed to gather it into his arms and proceeded down into the navigator-bombardier's compartment. In the nose section, the navigator, Lieutenant Carl H. Booth, Jr. and Staff Sergeant Donald L. Hissom, the togglier, were both trying to open the nose hatch, but it was stuck. Bush was informed by Booth and Hissom the hatch would not open.[10] The copilot, still struggling with his deployed parachute, now went back up into the cockpit.

In the meantime, both Booth and Hissom gave up trying to uncover the nose hatch, and decided to head for the bomb bay. As the two of them reached the cockpit, Bush was crawling through the top turret into the bomb bay. He stood there a moment, noticed the bombs still in the racks, and could see Hissom and Booth getting ready to join him. He then passed out from anoxia, and fell into the open bomb bay.[11]

The engineer, Technical Sergeant Clinton L. Bitton, exited the top turret, moved into the bomb bay, and jumped through the open doors.[12] In the tail section, Staff Sergeant George G. LeFebre, made it out the tail hatch, and his chute deployed successfully.

However, everyone in the waist section was having his problems. Staff Sergeant Thomas E. Therrien, the right waist gunner, assisted the wounded radio operator, Bennett, out the waist exit, and turned to aid the other waist gunner. He helped Fink with his parachute, and eased him towards the open waist doorway.[13] Weak from his bleeding wounds, Fink steadied himself to jump just as the plane rolled out of control. He passed out from the loss of blood, and fell out the door unable to open his chute.[14] At the same time, Sergeant Therrien was thrown backwards inside the aircraft. Therrien recovered, checked his chute, and once again moved towards the waist exit to jump. At that moment a 20mm shell exploded next to him, just inside the entrance.[15] He, like others that day, died flying his 25th mission.

The pilot finally was able to leave the disabled aircraft as he leaped through the bomb bay.[16] The bomber crashed at approximately 1350 hours near Waldenrath, Germany, some eleven miles inside the border.[17]

The Lang Crew

Many of the Fw 190s that had been vectored from far to the north down to the battle area had arrived during the early stages of the fighting, and were now departing to land, rearm, and refuel.

Their places were taken by swarms of single engine Bf 109s all sporting a new paint job, consisting of an orange colored nose and underside of the cowling, while the remainder of the fighter was painted black.

Their attacks against the crew of Lieutenant Robert S. Lang came from both the front and rear by individual, rather than two, four, and six sections of aircraft. Lieutenant Stanley Alukonis, Lang's copilot, who had been flying the aircraft for some time, gave the controls back to Lang at the pilot's request, and switched to the VHF radio frequency, while Lieutenant Lang monitored the intercom.[18] Alukonis took a good look back to his right, and suddenly realized Fisher's aircraft, that should have been abeam of Lang in the #2 position, was nowhere in sight.[19]

Alukonis looked to his right again and noticed a large hole in the #3 engine's metal cowling. A piston had just come through the top of it. The peeled back metal had opened up like the petals of a large flower. In order to keep the fighters from concentrating on his ship, Lang told his copilot not to shut down and feather the disabled #3 engine, but instead, let it run, and boost the power on the other three good engines.[20]

Another head-on Bf 109 attack caused damage to the left wing, and shattered a portion of the Plexiglas nose. Fortunately the bombardier, Lieutenant James G. Adcox and the navigator, Lieutenant John C. Tew, Jr. were not wounded.[21] Also at this time, Staff Sergeant Charles J. Groeinger, firing from left waist, reported a Bf 109 off at about seven o'clock had just smoked out![22]

The copilot looked across and over Lang, and out the pilot's side window. He could see the olive drab paint peeling off of #1 engine as the wing began turning red from an apparent internal fire. Suddenly the wing burst into flame as a fuel cell ignited.[23]

The pilot called over the intercom and ordered the crew to bail out. He then executed a steep, diving turn to the left to get down and away from the constant fighter attacks.

Lieutenant Alukonis did not hear the order to bail out, as he was monitoring the VHF channel. He suddenly noticed the engineer, Staff Sergeant Reuben B. Almquist, come out of the top turret, put on his chest pack, and enter the bomb bay. Lang reached to his right, and poked his copilot in the ribs to get his attention and then gave him a hitchhiking gesture with his free hand. Alukonis looked back at Lang with uncertainty. Lang poked him again, and once more he motioned to him it was time to go! This time the copilot got the message, slipped out of the right seat, and put on his chute. He stood on the flight deck for a few moments waiting to see what Lang was going to do.[24]

The pilot came out of his seat and slipped down into the nose section. The copilot followed. Lang took off his oxygen mask, and yelled to Adcox, the bombardier, and Tew, the navigator, "Bail out! Hit the silk, this is it!"[25]

By this time, Lieutenant Alukonis had released the nose escape hatch, and was preparing to bail out. He tried to go out feet first, but the slip stream swung his legs up against the side of the ship. He quickly pulled his legs back inside the aircraft and then pushed himself out the opening, head first.[26] The bombardier and navigator got ready to quickly follow. Jack Tew reached for his chest pack which had been sitting on his navigator's table, and he noticed it had been ripped and torn in several places from shell fragments.[27] He had not been wounded, but his chute was a mess. Without hesitating, he snapped it on and headed for the nose exit. Adcox followed. Just before Tew jumped, he glanced at his wrist watch – it was exactly 1348.[28]

In the meantime Sergeant Almquist, the engineer, had made his way aft through the bomb bay, past the radio compartment which was empty, and into the waist. He saw Technical Sergeant Warren E. McConnell, the radio operator, bending over the right waist gunner, Staff Sergeant Howard J. Keenan. He was reviving Keenan, who had passed out from lack of oxygen.[29] Almquist also noticed the groggy Keenan had set Murphy's Law into motion – his chest pack was on, but upside down. He steadied the shaky gunner, unbuckled the pack, turned it around, then snapped it back onto the harness in the correct position. He then led Keenan to the waist door, reminded him to pull the rip cord when in the clear, and pushed him out of the aircraft.[30]

The ball turret gunner, Staff Sergeant Kenneth A. Maynard, followed Keenan out, with McConnell right behind Maynard. The left waist gunner, Chuck Groeninger jumped immediately after McConnell, counted to ten, and opened his parachute.[31] Sergeant Almquist was now the last to leave the waist section as he too leaped through the open hatch.

Tail gunner Staff Sergeant Steve Krawczynski had been watching a Bf 109, which was on fire, when he heard the order to go. As he prepared to bail out the tail hatch, an exploding 20mm round nicked the little finger on his left hand.[32] Fortunately, the wound was only superficial. He looked toward the waist section and watched the exodus as the other gunners went out, one by one. He also noticed they had all deployed their parachutes almost as soon as they cleared the plane. As he now departed through the tail exit, he estimated the aircraft's altitude somewhere about 21,000 feet.[33]

At approximately 1352 hours, the Lang aircraft, still carrying its bomb load, crashed and exploded near the small German hamlet of Puffendorf. This village was about 14 miles northeast of Aachen and three miles southeast of where Holt's aircraft had gone in just two minutes earlier.[34]

The Kenyon Crew

After Lieutenant Dienhart left the formation at approximately 1349, Lieutenant Ellsworth H. Kenyon and his crew were all by themselves at about 21,000 feet – the only ship remaining from the low squadron. Most of the attacks this crew had experienced thus far had come from ten around through one o'clock, both from the high and level positions.[35] Fw 190s and Bf 109s had come at Kenyon's ship singly and in groups, and as they closed on the bomber they did a half roll and broke away underneath.[36] Now they also began attacking from the left and right sides.

The nose section and flight deck was still intact despite continuous head-on attacks, but the enemy fighters had done a job on the bomber's power plants. Three of the engines were now on fire.[37] Unfortunately, B-17 aircraft were only equipped to deal with a maximum of two engine fires, as just two carbon dioxide (CO_2) extinguishers were built into the system.[38] Kenyon now had to decide which of these CO_2 bottles to select for which two engines and at the same time hope the third fire would burn itself out.

More attacks came from both sides, and suddenly a 20mm shell exploded between the two waist gunners. Staff Sergeant Charles M. Green, the left waist gunner, went down with wounds to his right side and back. Technical Sergeant Richard W. Lewis, over at right waist, was more fortunate. His flak vest took the exploding fragments, and all he received was a blow in the back that felt similar to a kidney punch. He continued to fire at the attackers and observed some hits from his position on an Fw 190.[39]

The navigator, Lieutenant John A. Cole, observed his gunfire scoring some hits on a single-engine fighter, then his ammunition belt feed broke, and he had to stop firing and try to repair it.[40]

The damage to the aircraft was widespread. Kenyon now made a left turn to the southwest. He still was unable to put out the engine fires, and now someone reported a fire had broken out in the radio room. The aircraft was down to 20,000 feet as the pilot gave the order to bail out.[41] The ship was just approaching Duren, Germany, from the northeast as Kenyon tried to put some space between himself and the fighters.

Both waist gunners, the wounded Green and Dick Lewis, tried to leave through the waist exit but it was jammed and would not open. They began working on the jammed waist exit, trying to free it. It still would not open. Finally in desperation, lying on his back and using both feet, Lewis gave it a tremendous kick. The door came loose and so did Lewis' boots, and all three items disappeared out into the slipstream.[42] The two gunners quickly departed through the open doorway.

Staff Sergeant Walter L. Gottshall, still in the tail section, decided it was time to go. He released the hatch to his door, and was safely out.

Staff Sergeant Arthur Englehardt, down in the ball turret, exited up into the waist section, glancing at the fire behind him in the radio compartment. He caught sight of Technical Sergeant Russell R. Ahlgren lying on the floor, and the radio operator seemed to be wounded and was not moving. It appeared to Englehardt, Ahlgren had been hit in the head or shoulder.[43] The intense fire in the radio room kept Englehardt from getting any closer, so he headed for the waist door and jumped. Arthur Englehardt was another airman from Chelveston flying his 25th mission.

Up front there was a scramble to evacuate through the nose hatch. The engineer, Technical Sergeant Finley J. Mercer, Jr., slipped out of his top turret position, went down into the nose section, and left through that exit. Lieutenant Joseph F. Collins, Jr., the bombardier, flying today with Kenyon from the Eakle crew, immediately followed Mercer, and cleared the aircraft.

As Lieutenant Cole, the navigator, also prepared to use the same hatch, he saw the copilot, Lieutenant Thomas H. Davis, and his pilot, Kenyon, also coming down in the nose section in order to leave. All three got away safely in the vicinity of Duren.[44] At approximately 1353, Kenyon's B-17 crashed between Aachen and Duren, Germany, and the last ship of the low squadron was gone!

With the 305th low squadron annihilated, the German fighters could now concentrate even more fighters on the high squadron, or what was left of it. With Willis, Fisher, and Lang gone, it now consisted of Lieutenant Farrell and Lieutenant Robert A. Skerry. In actuality, there was no formation. Farrell and Skerry had started the day flying together, but they were now miles apart – Farrell out in front chasing Normand and Skerry far back trying to find some other B-17s to join.

Also at 1353, Lieutenant Colonel Milton, the 1st Division leader, who had led the bombers off course to Cologne now reached a point approximately 11 miles south of Bonn. Here he turned left to the southeast. After crossing back and forth several times over the Rhine while generally following the river's contour on his flight from Cologne to Bonn, Milton was finally back on course.[45]

However, for the 305th, the situation had now completely deteriorated since

Normand had chased after Milton's wing over the channel and left his high and low squadrons far behind. This was confirmed when Lang's ship from the high squadron crashed at 1352 while Normand was some 35 miles ahead of that aircraft. Further proof of this degradation occurred as Kenyon's low squadron B-17 went in at 1353, and his ship was 25 miles behind Major Normand's lead squadron.[46]

The Skerry Crew

When the enemy fighter attacks first began, Lieutenant Robert A. Skerry and his crew were far behind their scattered high squadron. They had never managed to recover from the turn over the channel and Normand's decision to accelerate and catch the wing in front of him. Lieutenant Willis, who had been flying on Skerry's right wing, stayed with his element leader as he, too, could not catch up to those ahead of him.

After the first assault on these two ships by the Fw 190s and Bf 109s, Willis' B-17 had staggered away to the left with considerable battle damage. Skerry had continued ahead, by himself, temporarily spared from the enemy's wrath, as he continued to follow after the 1st Combat Wing and his group.

At approximately 1356, Skerry skirted the edge of Cologne as he and his crew were being attacked repeatedly by Bf 109s, but so far they were holding their own.[47] The Bf 109s concentrated mainly on head-on attacks from ten o'clock around to the two o'clock position. They came singly and in pairs.[48]

Skerry now observed a twin-engine Bf 110 making a pass at his bomber from the left side. The German pilot let go with a pair of rockets, and fortunately he missed.[49]

Several Fw 190s now joined the battle making their firing passes from the tail and both sides. One of these German pilots was *Oberfeldwebel* (1st Sergeant) Willi Roth from the 5./JG 2.[50]

Technical Sergeant Edwin E. DeVaul, the tail gunner, reported from the rear a good portion of the vertical fin and rudder had disappeared. In addition, he informed his pilot there were numerous holes in the horizontal stabilizer and right and left elevators.[51]

Up front, number one engine on the left side began losing oil pressure, had to be shut down, and feathered.[52] Just moments later, several rounds of a 20mm cannon exploded in the radio room and left a large, waist high hole in the compartment door.[53]

The radio operator, Technical Sergeant Stanley H. Larrick, broke in on the intercom, "This is the radio man, I think that my leg is blown off!"[54]

Tail gunner DeVaul again heard Larrick's voice come on the intercom. The radio operator kept repeating he had been hit and his leg had been blown off – then his voice trailed off and the intercom became silent.[55] The bomber was under such continuous heavy attack no one was immediately available to leave their position and go to assist Larrick.

Skerry turned south towards Bonn. He now passed over onto the east side of the Rhine river.[56] At 1357 hours, he was immediately taken under fire by an 88mm German flak battery stationed southeast of Cologne. The battery fired for one minute and got off 20 rounds at the bomber.[57] The Germans observed the B-17 losing altitude, but then lost visual contact with it due to the dense artificial fog and smoke being dispensed to hide the area.[58] As the ship continued south, a 105mm battery joined in, and fired 38 shells at the bomber.[59]

At 1359 Skerry crossed back over the Rhine to the west side and approached Bonn. Three 83.5mm batteries took over from the 105s, and between them they expended approximately 50 more projectiles at the American bomber.[60] The antiaircraft units estimated his altitude at approximately 23,000 feet, and the aircraft received considerable damage from some of the accurate ground fire. Another engine, this time on the right side, lost power and had to be feathered.[61] The copilot, Lieutenant John C. Lindquist, observed a large hole in the right wing which had not been there just moments before.[62] *Oberfeldwebel* Roth, the Fw 190 pilot who was to be credited with shooting down Skerry's aircraft, continued to attack from the six o'clock position scoring numerous hits on the reeling bomber.[63]

The bomber was now down to 15,000 feet, and the pilot could no longer maintain altitude. He decided it was time to go, gave the order to bail out, and hit the bail out bell.[64] Immediately after that the intercom became inoperative.[65]

Down in the nose section, Lieutenant Robert Guarini, the navigator, reached for his parachute and, realizing the intercom was dead, motioned to Sergeant Cecil S. Key, the togglier, it was time to abandon ship. Key, flying his first mission, just sat on the floor and did not move. Lieutenant Guarini, suffering from wounds to the thigh and head, again gestured to the togglier they had to bail out. Sergeant Key continued to sit. He seemed frozen and could not move. The navigator gave it one more try, but he could not budge Key from his position.[66]

Staff Sergeant Robert J. Middleby, the engineer, left the top turret, moved down into the nose section, and bailed out through the nose hatch. He was followed by Lindquist, the copilot, and then Guarini.[67] Just before he exited the ship, Bob Guarini glanced back at Key. The gunner still had not moved.[68]

Staff Sergeant Wayne D. Rowlett, the ball turret gunner, rolled his guns down,

Approaching the Rhine

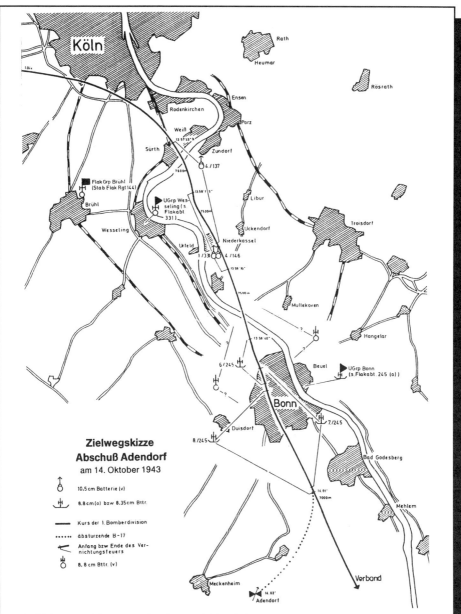

The flight path of the 1st Bombardment Division (minus) as it passed over the southern edge of Cologne (Köln), turned sharply to the south, and flew over Bonn. The bottom of the page shows the Skerry aircraft crash site at Aldendorf at 1402 hours. (Map courtesy of Friedhelm Golücke and Verlag Ferdinand Schöningh. Taken from *Friedhelm Golücke: Schweinfurt und der strategische Luftkrieg 1943*, Paderborn 1980 p. 256)

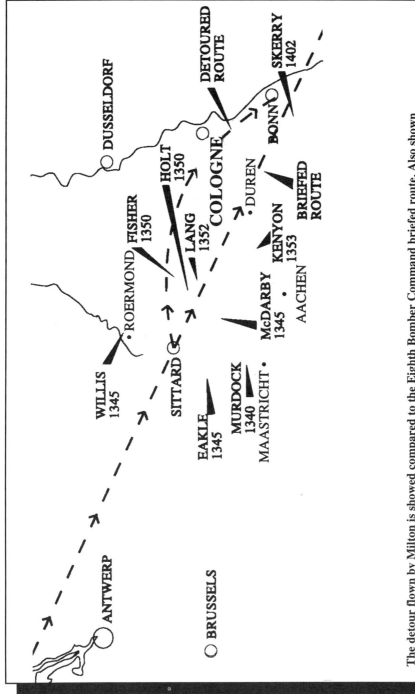

The detour flown by Milton is showed compared to the Eighth Bomber Command briefed route. Also shown are the first nine 305th Bomb Group crash sites. The tenth ship, flown by Dienhart, was missing and unaccounted for at this time.

and exited into the waist section. As he pulled himself from his turret, he glanced towards the radio room and saw the radio operator was already dead.[69] Sergeant Larrick was lying on the floor of the radio room in a pool of blood with one leg almost completely off![70]

Rowlett approached both waist gunners, Staff Sergeants Jack G. Johnson and Gus Doumis, and all three prepared to bail out. As Rowlett stepped to the waist exit an explosion rocked the waist section. Wayne Rowlett was blown out the doorway, while Sergeants Johnson and Doumis were knocked back into the waist.[71]

Gus Doumis got to his feet and reached the doorway. He turned and quickly glanced back at Johnson. He saw Johnson return to a position between his right waist gun and the waist escape hatch. The ship was hit again, and Doumis saw Jack Johnson double up and fall to the floor of the aircraft as though he had been hit in the stomach. The next shell burst threw Doumis out of the waist hatch, and he did not see if Johnson managed to leave the ship or not.[72]

When the order came to bail out, Sergeant DeVaul, the tail gunner left his seat, and retrieved his chute which was near the tail escape hatch. He looked forward and saw Sergeant Jack Johnson, the right waist gunner, with Sergeant Gus Doumis, the left waist gunner, returning to their positions with their chutes. DeVaul now planned to go to the aid of the wounded radio operator, but as he began to crawl forward towards the waist section, a 20mm projectile exploded nearby hitting him in the lower back and buttock. As the wounded DeVaul pulled himself over to the tail exit to leave, he quickly scanned the entire waist section and saw it was empty. He was the last one to depart from the rear of the aircraft.[73]

Bob Skerry set the automatic pilot and waited a short time for the crew to clear the aircraft. He was the last one out as he left by the nose exit.[74]

The bomber crashed and burned 10 miles southwest of Bonn at 1402 hours.[75] With the crash of the Skerry aircraft, the high squadron, less the Farrell crew, was gone.

From 1340 hours, when the Murdock aircraft slammed into the ground, until 1402 when Skerry's B-17 crashed, Major Normand, leader of the 305th, had lost two-thirds of his command. In just 22 minutes, 10 of the 15 B-17s from the 305th Bomb Group had fallen to the enemy. The entire six ships, of the low 364th Bombardment Squadron, sent on the "Second Schweinfurt" raid not only did not reach the target, but they never got past the Rhine river. In addition, four planes, from the high 366th squadron, suffered the same fate.

From where these 10 aircraft crossed the Dutch coast entering the European continent, nine of them were only able to get approximately half way to the target,

Schweinfurt, before they were destroyed. The tenth B-17, Dienhart's ship, was unaccounted for at this time. Major Normand, and what was left of his outfit, still had some 115 miles to fly before reaching Schweinfurt.

Chapter 10

The Bomb Run Before and After

Before

After Lieutenant Colonel Milton brought the meandering 1st Division (minus) back on track from its detour to the northeast, he had returned to a southeasterly heading of 120 degrees.[1] The fighting continued uninterrupted as Milton now adjusted his heading to the left to about 102 degrees. He now proceeded to the next check point which was 16 miles north of Frankfurt. The bombers would arrive there in 24 minutes and more fighters would be waiting.[2]

The detour taken by Milton northeast to the Cologne area caused further confusion for the 1st Division. It cost Milton some time and distance, and he was no longer out in front. Colonel Peaslee's 40th Combat Wing, which was behind Milton since the leadership change over the channel, was now on Milton's left and four miles ahead of him.[3] Peaslee had stayed on course and not followed Milton as he misled the division in the wrong direction. Lieutenant Colonel Rohr's 41st Combat Wing, which also had stayed on course, was not affected as it continued on course – now about 10 miles behind Milton's wing.[4]

The savage fighting continued with no let up as twin-engine day fighters from their base at Wertheim, 15 miles southwest of Würzburg, joined the fighting and began attacking. Many of them concentrated on what was left of Peaslee's wing, as the 40th Combat Wing continued to be under heavy pressure from the swarming enemy.[5]

Unlike the 40th, the 1st Combat Wing, though under some attack, was having a better time of it. In the meantime, Major Normand and his lead squadron were still clinging to the low, left position in Lieutenant Colonel Milton's 1st Combat Wing.

The leader of the 305th now had a bomb group consisting of four battered and battle weary aircraft – all from the lead squadron. Besides himself, Normand still had Kincaid on his right wing, Bullock off his left side, and Maxwell just behind and below him. Despite the ferocious German fighter attacks, these four aircraft had managed to avoid any major battle damage, were still airborne, and fighting their way to the target.

Other than the two earlier abortions by Lieutenants Reid and Chely prior to the start of the air battle, Normand had not lost any B-17s from his original lead squadron to the enemy. This was due to several reasons. First, the ships in his squadron were close to his and had been able to respond quicker to Normand's erratic decisions, and second, the 1st Combat Wing, to which Normand had attached his squadron, had flown by the book. Smooth shallow turns, and gradual increases in acceleration and deceleration had all been combined to enable the pilots with the 1st Combat Wing to maintain a close, intact formation.[6]

However, this had not been the case for the dispersed B-17 formations of the 305th's high and low squadrons, which left far behind and strung out by their group leader, had been simply overwhelmed by the large number of enemy fighters. With the exception of the Farrell crew, the others had been picked off, one by one.

At 1410, many German fighters were told by ground control the target for attack is probably Frankfurt – but it was wrong.[7] Milton, the designated 1st Division leader, continued on and at 1417 reached his checkpoint located north of Frankfurt. The fighters, who were sent to the Frankfurt area earlier, now joined the action. Peaslee, still off to the left of Milton, had already past this point two minutes earlier and was now seven miles ahead of Milton. At this rate, Peaslee would bomb first.[8]

Lieutenant Colonel Milton now turned further southeast to a heading of 120 degrees and flew towards Würzburg, the initial point (IP).[9] The bombers still had some 64 miles to fly before reaching this city, where they would turn and make their bomb run to Schweinfurt.

There was another bomber from the 305th soon to join Normand. It was that of Lieutenant Farrell the last ship remaining from the high squadron. Farrell, still chasing Normand, was now closing the gap to the lead squadron and was not too far behind.

The Farrell Crew

Lieutenant Frederick B. Farrell had begun the day leading the 366th high squadron of the 305th Group. Major Normand's earlier decision to circle over the channel, then accelerate leaving the high and low squadrons chasing to catch up, had whittled Farrell's outfit to one B-17, his. At approximately 1413, still pursuing Normand and four minutes from reaching the Frankfurt checkpoint, Lieutenant Roy A. Burton, Farrell's copilot, confirmed a "kill" for his engineer. Up in the top turret, Technical

Sergeant Marvin D. Shaull had just blown the tail off a passing Fw 190.[10]

At right waist Staff Sergeant Jayson C. Smart observed several German fighters in trouble. He saw one explode and another began to smoke.[11]

Just after departing the Frankfurt area, the Farrell crew received several attacks from six o'clock level. Immediately afterwards, the tail gun position became strangely silent. Right waist gunner Smart realized something was wrong back there and called his buddy, Staff Sergeant Thaddeus J. Niemiec, on the intercom but received no answer. Since he was not busy at the time, Smart decided to go to the aid of his friend. Just as he started to go aft, he saw Niemiec crawling around the tail wheel strut towards him. Niemiec jumped to his feet and rushed to Smart.

The waist gunner pulled off his oxygen mask and shouted, "You all right?"

The tail gunner yelled back, "Yeah, I need more ammo, I'm all out!"[12]

With that, Smart returned to the right waist, and muttered a prayer of thanks to the Almighty – his buddy was okay. Niemiec went forward to the radio room, picked up more ammunition, then double-timed back to his tail position. The gunner arrived back in his spot none too soon. Several Bf 110 twin-engine fighters had pulled in behind the bomber and were maneuvering into position to fire their rockets at the B-17. Sergeant Niemiec fired, and the Germans fired – both missed.[13]

Lieutenant Barney Farrell and his crew continued to draw closer to Normand's lead squadron and at approximately 1422, 10 minutes from the IP, he finally caught up to Normand.[14] As he approached the lead squadron, Farrell slid his aircraft into a position above and on the right of Lieutenant Kincaid, who was Normand's right wingman.[15] Normand and the 305th Group now had a five ship formation.

Visibility continued to be unlimited. At approximately 1423, Lieutenant Joseph Pellegrini, the group lead bombardier for the 305th flying with Major Normand, saw the IP when the bombers were still 35 miles northwest of the city.[16]

The bitter fighting kept on uninterrupted as the German planes continued to press the 1st Division bombers with head-on and attacks from the tail. With few stragglers left to destroy, the record number of attacking German fighters looked for any small remaining elements to attack as they avoided the larger, tighter formations whenever possible. Many of the enemy now bounced what was left of the five struggling 305th B-17s.

The Maxwell Crew

At 1427 hours, approximately five minutes northwest of Würzburg, Lieutenant

Victor C. Maxwell's ship was flying at 21,700 feet in the #4 position behind and just below Normand.[17]

Technical Sergeant Silas W. Adamson, the engineer, claimed knocking down an Fw 190 from his top turret position as he watched the fighter smoke and parts fly off of it. Moments later he claimed another but could not follow the fighter as it disappeared from sight.[18] Down in the nose section the navigator, Lieutenant Urban H. Klister, was wounded by shell fragments from several exploding projectiles.

Lieutenant Andrew J. Zavar, the bombardier, called over to the navigator on the intercom, "Lieutenant Klister, do you want the first aid kit?"[19] The navigator was last heard refusing any medical aid.

The fighting continued. Maxwell managed to hold his position in the formation despite the battle damage his ship was receiving from the fighter attacks.

Now three minutes from Würzburg, the Maxwell crew became the object of a concentrated fighter attack from the 11 o'clock position. His aircraft was rocked by numerous explosions, fire broke out in the forward portion of the plane, and the bomber began to spin.[20] After losing several hundred feet, the pilot managed to stop the rotation of the aircraft, leveled off, and gave the order to bail out.

Staff Sergeant William H. Connelley, the tail gunner, went out the tail escape hatch, cleared the ship, and popped his parachute. He estimated his altitude at 20,000 feet.[21] The left waist gunner, Staff Sergeant Herbert S. Whitehead, together with Sergeant Charles H. Crane, the right waist gunner, grabbed their chutes and headed for the waist exit. Whitehead jumped clear. Just prior to his leaving, Crane turned and saw Technical Sergeant Jerome B. Pumo, the radio operator, closing the radio compartment door and getting ready to leave.[22]

At that moment, 1429 hours, Staff Sergeant John F. Raines, facing aft in the top turret of the Kincaid B-17, which was flying on Normand's right wing, glanced back and down to his right. Just as he looked at the stricken Maxwell aircraft, it exploded and disintegrated in midair.[23] It simply disappeared in a large ball of fire and smoke – it was there one moment and gone the next! Technical Sergeant Carl J. Brunswick, in the top turret of the Bullock aircraft, which was on Normand's left wing, also saw the Maxwell ship and its crew disappear as they blew up in a cloud of debris.[24]

The unlimited visibility continued to dominate the target area, for at 1429, when he was 12 miles from Würzburg, Normand's bombardier, Pellegrini, looked off to his left and could clearly see Schweinfurt some 23 miles away.[25]

At 1430, two minutes shy of reaching the IP, two twin-engine Bf 110s came at the Farrell aircraft in a coordinated rocket attack. One approached from three o'clock level while the other came up from five o'clock low. Both fighters missed, and

Farrell's gang, flying a ship named "Rigor Mortis", had been in the right place at the right time – again.[26]

The incessant, wild fighting continued and moments later Lieutenant Farrell eased his ship away from Kincaid's right wing. He now dropped down and behind into the #4 slot that had been Maxwell's.

Lieutenant Colonel Milton, leading with the 91st Group, called the other two groups in his 1st Combat Wing just prior to reaching the IP. He told the 381st and 351st he planned to turn a little short of the IP. This would give Peaslee, who had slowed, an opportunity to get in trail for the bombing run. Milton also reminded his groups not to take too much interval for their bombing runs and continue to keep their formations tight.[27]

So far the lead 1st Combat Wing had lost only one aircraft since the battle began. Not including the 305th, Milton's box still had 32 bombers. This was an incredible accomplishment against an overwhelming enemy![28]

Peaslee's 40th Combat Wing, including the absent 305th, had 23 B-17s left out of 49. His three groups had lost 26 planes and had yet to reach the target.[29]

Lieutenant Colonel Rohr, coming along behind Peaslee and leading the 41st Combat Wing, counted 48 heavies still with his three groups. Like Milton, he also had only lost one aircraft to this point, another extraordinary feat![30]

It was obvious, up until now, the Luftwaffe had been feasting on the three groups of the dispersed 40th Combat Wing. These were the units that were required to do all the controversial maneuvering and position changing over the English channel – the 92nd, 306th, and 305th.

Colonel Peaslee, who originally led, then trailed, then moved ahead again when Milton drifted off course and overflew the Cologne and Bonn areas, now was barely behind Milton as the turns at the initial point were completed. Milton arrived at the IP at 1432 and Peaslee at 1433.[31]

As the leader of the 1st Division passed over the initial point, he began a shallow turn to the left to a heading of 44 degrees. The target, Schweinfurt, was some 19 miles ahead, but it could be seen clearly with the excellent visibility that existed. As Colonel Milton's aircraft rolled level at 22,300 feet, his bomb bays door opened, and everyone still with him followed suit. The bomb run had begun – it would take eight minutes.[32]

Due to the unlimited visibility, Lieutenant Joseph Pellegrini, Normand's bombardier, was able to pick up his actual aiming point while over the IP. As Normand's aircraft completed the turn at the IP, Pellegrini called Normand on intercom and told him he was beginning the bomb run. He asked Normand to confirm the automatic

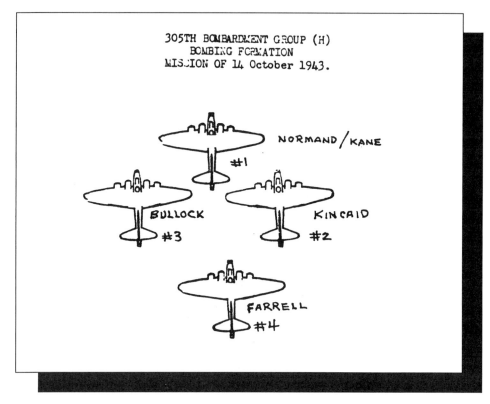

The formation of the remaining four ships of the 305th Bomb Group as it began the bomb run. (USAF Historical Research Agency)

pilot (AFCE) was engaged.[33] Pellegrini needed to integrate the AFCE and the bomb sight to provide a rigidly controlled platform from which to work. Using this procedure, the bombardier controlled the heading, turns, and drift while the pilot maintained the correct airspeed and altitude. Using the AFCE together with the Norden bombsight afforded a bombardier the best possible results.

Normand's response was that of doubt. He told Pellegrini the AFCE had been damaged from an earlier cockpit explosion.[34] Pellegrini came back on the intercom and recommended they use the pilot's directional indicator (PDI). With the AFCE off, Pellegrini could make a manual run utilizing the PDI. The pilot simply followed a needle right or left and maintained airspeed and altitude, as the bombardier fed small drift corrections into the bombsightt.

Major Normand told Pellegrini he was not sure the PDI was working either, and also advised his bombardier, due to having only four ships left in the group, it would be better to attach themselves to the group ahead (Milton's) and bomb with them.[35]

Lieutenant Pellegrini persisted. "I told him, by God Major, I can see the target in my bomb sight. I can see what we came all this way to hit. I said I knew I could hit it, but he didn't think we could hold it on the run . . ."[36]

The aircraft continued on the heading for the 305th's aiming point. Pellegrini, who had already set the cross hairs of the bombsight on it, continued hunched over the aiming device making final adjustments and corrections for the 305th's bomb run.

Pellegrini watched through his bombsight eyepiece as the 305th target moved completely out of sight. The 305th now pulled in closer to Milton's 91st Group on its right. Normand came back on the intercom and told Pellegrini to drop on the 91st Group's bombs.[37]

Pellegrini called and literally begged Normand, "Major, please let me continue this run. Visibility is perfect, we can't miss!"[38]

Normand's reply was to drop on the bombs of the group to their right. Lieutenant Pellegrini could not believe what he had just been ordered to do. Disgustedly, he raised up from over the Norden bombsight, turned off the electrical power to the unit, caged its gyros, and slumped in his chair. After all it had been through, what was left of the 305th Bomb Group was denied its own bomb run, and Lieutenant Pellegrini was ordered to drop his bombs when he saw the 91st Group release theirs.[39]

The lead unit of the 1st Division, the 1st Combat wing, was still receiving some fighter attacks as the bomb run continued, but the enemy now seemed to be concentrating on the 40th and 41st Wings, which continued to follow Milton.

The flak would start as soon as the bombers came within antiaircraft weapon range of the target. The fighter attacks would then slow or stop temporarily, as the ground units would take over and try to increase the total of downed bombers. After the "heavies" moved out of range, the flak would end, and the fighters would go to work again.

The Kincaid Crew

As the 305th made the turn over the IP at 1432, Lieutenant Alden C. Kincaid, on Normand's right wing, had been flying with a dead copilot for over an hour.[40] Lieutenant Norman W. Smith, the young man from Hawaii, had been killed in the first few minutes of the early fighting when the bombers had moved across the Dutch border into German airspace.

Machine gun fire, from a head-on attack by an Fw 190, had raked the cockpit wounding Kincaid in the right arm. Smith was struck by a round that passed through his left forearm and into his side killing him instantly. Staff Sergeant John F. Raines, the engineer, called the navigator, Lieutenant Robert D. Metcalf, and had him come up from the nose to help with first aid.[41] Even though Raines was directly behind the pilot and copilot positions in the top turret, he could not leave his guns due to the swarming fighters. Metcalf wrapped Kincaid's wound, and finally got it to stop bleeding. When he turned to help Lieutenant Smith, who was slumped far to the right in his seat, there was nothing he could do for him – he was gone.[42] Ironically, the young man, who wanted to fight the Japanese in the Pacific, died at the other end of the war – over Germany.

Down in the ball turret Staff Sergeant Kenneth R. Fenn and the pilot of a Bf 109 were playing a game of cat-and-mouse. Each thought the other was the mouse. The fighter was approaching from 12 o'clock low, and Fenn had been following him for a mile. The fighter was flying a zigzag course, as if to indicate he was breaking off the attack and leaving. Fenn would not buy it. The fighter's nose always came back pointing directly at Kincaid's ship. Suddenly, the turning and changing of direction stopped, and the fighter charged the B-17. Sergeant Fenn's patience finally paid off. As the Bf 109 came in range, Fenn started his turret's "twin fifties", and followed the enemy all the way to within several hundred yards of the bomber. Parts, including the propeller spinner, began coming off the fighter, and heavy black smoke trailed from the engine. The Bf 109 fell away, and spun out of sight.[43]

At 1435, Sergeant Louis Bridda, the left waist gunner, had just returned to his station from the nose section. Lieutenant Phillip A. Blasig, the bombardier, had expended all the ammunition he brought along for his nose gun, and Bridda had given him what he could.[44] They were still five minutes from the target, and .50 caliber ammo was getting scarce.

Bridda now saw a passing Fw 190, as it began to smoke and break away. From his position, he could see three more fighters in trouble, as other gunners began scoring.[45]

Back in the tail section, Sergeant William C. Heritage was busily engaging a pair of Fw 190s displaying black and white checkerboard squares on their noses.[46] Ball turret gunner Fenn picked up a Ju 88 at seven o'clock low just as the twin-engine fighter let go a pair of rockets. They missed, and Fenn's tracers chased off the German pilot.[47]

The Kincaid aircraft now took several direct hits from an incoming fighter that completely wiped out the pilot's controls and set #3 engine on fire.[48] In addition, #4 engine began to lose power.⁴⁹ Raines, up in the top turret, saw #3 engine over on the right burning, and swung his turret around to check the other side. There he saw some six to eight feet of the left wing missing.⁵⁰

Back in the waist section, Lou Bridda called Fenn on intercom and told him, "Okay ball turret, let's get the hell out of here!"⁵¹ He had called just in time, as moments later the intercom was shot out.⁵²

Kincaid tried moving the controls in various directions, but they were loose and completely unresponsive. He decided it was time to go. The pilot tried the intercom and it did not work, so he rang the bail out bell and passed the word, as best he could, to the crew to bail out.⁵³

Heritage, from his tail gunner position, never heard the alarm bell to bail out. However, he could see #4 engine was disabled and a large section of the left wing was now missing. The ship was at approximately 21,500 feet altitude when he left through the tail exit.⁵⁴

It was 1437 hours as the aircraft fell off on the damaged left wing and slowly began to spin. Inside there was a mad scramble by the crew to get out of the ship. Many of them had trouble moving towards the exits as the centrifugal force caused by the twisting B-17 pinned them down inside the plane.⁵⁵

Sergeant Alfred C. Chalker, who had been firing from the right waist aperture, felt a tap on the shoulder, and turned to see Bridda pull the quick release handle for the waist door exit. Then he watched as Bridda went out the door. Chalker got the message. He put on his chute, but was immediately thrown to the floor and could not move.⁵⁶

Sergeant Fenn managed to clear the ball turret, put on his chute, and get within a few feet of the waist exit just before the ship started spinning. He, too, was now pinned down on the floor next to Al Chalker. Fenn kept struggling, but could not overcome the force holding him down. Finally he quit trying, gave up struggling, and resolved himself to the fact, "this was it"! He could see out the open waist exit, and the ground was coming up fast. He decided to try to get out one more time. He managed to get alongside Chalker and literally threw his friend out of the aircraft. Then he somehow dragged himself out the door and immediately pulled the rip cord – he was about 2000 feet above the ground.⁵⁷

Al Kincaid, the pilot, was able to crawl down into the nose section despite the rotating aircraft. Lieutenant Metcalf had already left, and Lieutenant Blasig was attempting to get to the nose exit. Kincaid went out followed by Blasig.⁵⁸

The engineer, John Raines, was able to get out of his top turret and snap on his chute. He had trouble moving, but managed to crawl into the cockpit, roll from the flight deck forward, and then down into the nose compartment. From there, he somehow managed to go feet first out through the open nose hatch.[59]

At 1439, the Kincaid aircraft crashed approximately one mile southwest of Schweinfurt just a minute from the target. So near and yet so far![60]

With the Kincaid aircraft gone, Lieutenant Farrell, who had recently occupied the #4 position after the Maxwell tragedy, now maneuvered his ship right, up, and forward and took over the spot vacated by Kincaid. Farrell now occupied his fourth different position of the day. He took over the #2 position as right wingman for Normand, opposite Lieutenant Bullock's plane, which was in the #3 spot. These three B-17s were now all that remained of the 305th Bomb Group.

In comparison to what had already occurred to their comrades in the other 12 bombers, this trio had experienced a much better day of it – up to this point in time. These three airplanes had picked up considerable battle damage along the way, with numerous holes appearing in the wings and fuselage. As the bombers were about to reach the target, each one had all four engines still functioning, and there were no dead or wounded on board. However, Bullock had a problem – his left wing was on fire!

The Bullock Crew

All day long those on Lieutenant Raymond P. Bullock's crew had been leading charmed lives. For four of them, they just needed to get through this one more mission, their 25th, and for them their combat tour was over.[61]

Right waist gunner, Staff Sergeant Stanley J. Jarosynski, had a weapon problem earlier in the fighting when his gun jammed. However, in a short period of time, he was able to clear it and continued firing. Both Jarosynski and Staff Sergeant Harold E. Coyne over at left waist, had not been getting much to shoot at throughout the day, as most attacks had come at them from the front or rear.[62]

The same was true for Staff Sergeant Joseph K. Kocher. Down in the ball turret, Kocher felt more like a spectator than a participant. Just as the two waist gunners he, too, had very few shots at any fighters for the same reasons. He did notice parts fly off an enemy fighter as it took a pounding from somewhere. He also saw another smoke and fall away. Again, he did not know who scored. Another grim observation by Kocher – he saw more B-17s go down than fighters.[63]

Unlike the other three gunners who did not have many targets to take under fire, Technical Sergeant Carl J. Brunswick, up in the top turret looking fore and aft, saw numerous single-engine fighters attacking from the 12 o'clock level position. In addition, he observed twin-engine fighters firing rockets into the rear of several formations from the six o'clock position. He also saw a number of his rounds accounting for some fighter damage, but could not tell if the enemy went down or not. On one occasion he ran out of ammunition, but fortunately he had stockpiled more on the floor by the turret.[64]

Up until the IP, this crew's ship had some holes in it, but had been spared any major battle damage. At 1434 hours, two minutes after passing the IP while still six minutes from the target, this all changed! At this time, Bullock's aircraft took a 20mm round into the left wing fuel cell. It exploded just outboard of #1 engine, and raw gasoline began to pour from the trailing edge of the wing. Moments later the wing began to smoke. Lieutenant Bullock glanced to his left and could see the smoke had turned to fire and now most of the wing was burning.[65] The B-17 was handling just fine, but the fire was getting worse.

Bullock was still some six minutes from "bombs away", and he had a quick decision to make – should they bail out or stay in formation and try to bomb the target. The heat from the fire inside the wing could cause it to burn off any time, and the aircraft was liable to explode from the burning fuel. Lieutenant Bullock now made a courageous decision. He called the crew on intercom, explained the situation, and told them to bail out if they wished. He declared he was going to finish the "run", and complete what they came there to do – bomb Schweinfurt – then he was leaving. They all stayed.[66] Bullock and his crew members were willing to pay the supreme sacrifice so they could complete the bomb run with their group.

Unknown to the men of the Bullock crew their group leader, Major Normand, had other ideas and made yet another questionable decision. He ruled out a 305th bomb run!

What really took place as the 305th prepared to bomb? In his book, *Decision Over Schweinfurt*, Thomas M. Coffey wrote, "On Milton's left, the 305th Group was pushing bravely forward under command of Major Normand though twelve of its fifteen planes had already fallen and only three remained in formation. After looking at this pitiful remnant, Milton picked up his microphone and asked Normand if the 305th would like to pull in and make its bomb run with the 91st.

Normand answered proudly, 'No, sir. The 305th will make its own bomb run.'"[67]

If this conversation did take place between Milton and Normand, then Normand's reply was a pretense on his part. Lieutenant Joseph W. Kane and Lieutenant Joseph Pellegrini, Normand's pilot and bombardier, both agreed Normand would not allow a group bomb run, but instead decided the 305th's bombs would be "toggled" on

Milton's 91st Group which was on Normand's right.[68]

Their statements are further substantiated by the 305th's Intelligence Officer's report, published later that day which contained the following information from the S-7 section, "Our lead bombardier was able to synchronize his bombsight directly on the MPI [mean point of impact]. He called for a separate bomb run not knowing that at the time there were only two other A/C [aircraft] in our formation. Our group Commander [Normand] decided wisely to continue to fly with the group ahead as both groups were under E/A [enemy aircraft] attack at the time . . ."[69]

The 305th group operations officer's report, filed the day after the mission, also contradicts the reported radio conversation between Milton and Normand. It read, ". . . As only three aircraft remained of the 305th group formation the 305th bombed the center of the city instead of the briefed aiming point [AP]. (The bombing of the briefed AP would have caused the three planes of the 305th group to become separated from the other groups)."[70]

Coffey also wrote, "The three remaining planes in the 305th Group, still under attack by fighters, followed grimly behind the 91st, but as they approached the target, Lt. John [Joseph] Pellegrini, the lead bombardier, expressed dissatisfaction with their position. Convinced they were slightly off target, he suggested they go around and make a new approach. Normand, aware that if they took time to do so they would be separated from the rest of the 1st CBW, told Pellegrini to do the best he could. As a result, the three planes of the 305th dropped their bombs on the center of the city"[71]

Kane and Pellegrini have also agreed Pellegrini did not suggest to Major Normand the 305th make a 360 degree turn and go around for a second bomb run.[72] Normand, who had the final say, had already made the decision he would *not* make a group bomb run, but instead the 305th would follow the 91st group and bomb with them. So Kane and Pellegrini did as ordered. Kane flew close to the 91st Group, and Pellegrini toggled on Milton's bombs. At the 1st Division commander's meeting the day after the mission (chapter 12), Normand further damaged his credibility by telling his superiors – it was Pellegrini who decided against making a *305th* bomb run![73]

The bomb run continued. The 305th followed alongside Milton's 91st Bomb Group. The Kane aircraft and the one flown by Farrell were still intact. The left wing of Bullock's ship continued to smoke and burn.

The Kane Crew

The 305th's group lead B-17 flown by Major Normand's pilot, Lieutenant Joseph W.

Kane, was also filled with holes from both flak and machine gun fire, but so far the crew had been spared any serious injury to themselves or major battle damage to the aircraft.

Earlier in the day, shortly after Kane's bomber entered German airspace, a yellow-nosed Bf 109, making an attack from 12 o'clock high, put two rounds of 20mm behind the pilot's instrument panel. One detonated and blew out many of the instruments and parts of Lieutenant Kane's overhead and front window. His helmet was burned, his hair singed, and he received some minor facial cuts from flying Plexiglas. Major Charles G.Y. Normand, over in the right seat, had been untouched by the exploding shell.[74] Unknown to both of them, the other partially unexploded 20mm projectile was imbedded behind the instrument panel and could go off at any time.

At 1438 the visibility was clear and unlimited as Milton's 1st Combat Wing and the 305th were two minutes from "bombs away." As the bombers approached the target, the fighter attacks diminished and black bursts of flak began to appear. The sky around the remaining three 305th bombers was a mass of bursting flak as they began receiving intense, accurate antiaircraft ground fire.[75] For some of the other groups, such as Peaslee's 92nd, the flak was moderate and fairly accurate but his group was not bothered by it.[76] For Lieutenant Colonel Rohr's 41st Combat Wing, the flak was very light.[77]

Moments before "bombs away" most of the fighters were gone, but a few continued to attack the formations. They flew through their own flak to get at the bombers.

In the meantime, many of the German fighters not in the battle were having problems of their own – they could not locate the bomber formations. As the 1st Division prepared to unload on the target, two squadrons of twin-engine day fighters from the Quackenbrück sector reported they were landing to refuel without any contact with the enemy. Also at this time, after failing to locate the B-17s, two single-engine squadrons from Bremen landed in the Bonn region. Despite the setbacks, the German fighter command continued to keep the pressure on as several more formations of twin-engine night fighters, from west of Berlin, continued their flight south in the hopes of engaging the bombers.[78]

Peaslee's 92nd Group was practically on top of one of the 1st Combat Wing units when they arrived over the target. At 23,800 feet the 381st Group, which was just up ahead in Milton's outfit, bombed first at 1439. Seconds later Peaslee's navigator recorded "bombs away" at 1439 and one-half from 23,000 feet. These two were followed at 1440 hours by Milton's 91st Group which was at 23,100 feet. The 305th

Group at 21,600 feet dropped on the falling bombs of the 91st also at 1440. The remaining units of the 1st Division continued their runs, and it was all over at 1446 when the 303rd group from the 41st Combat Wing finished the job from 24,050 feet.[79]

Ten of the 18 ships that took off with the 305th Group in late morning carried cameras for taking pictures of the results of the bomb run. One of the ships that aborted, the Reid crew, took its camera home, while eight more of them were destroyed when their aircraft were shot down. When Pellegrini toggled his bombs with the 91st Group, the Farrell aircraft had the only 305th camera remaining at "bombs away" – so he was able to take the pictures.[80]

After

As the 1st Division came off the target, it was now attacked by a large number of rocket firing twin-engine fighters with 16 of them concentrating on the 41st Combat

Smoke bombs falling from a 305th group leader indicates bombs away on an enemy target. The other ships in the formation then toggled their bombs with those of the group leader. (305th Bomb Group photo)

Wing which was last off the target.[81] The flak stopped and other fighters now flocked to the bombers.

Two of the three remaining B-17s of the 305th Group, Kane's and Farrell's, followed the 1st Combat Wing in a diving right turn off the target in order to begin their long return flight home. The other one, Bullock's aircraft, continued straight ahead but not for long. The crew had already made preparations to bail out – all they needed now was the word to leave.

Immediately after "bombs away", the order came to abandon ship. Lieutenant Homer L. Hocker, the copilot, passed the word over the intercom while Bullock reached to his left front and flipped on the bail out bell.[82] It was time to go. The entire left wing of the plane was now engulfed in flames, and the fire inside that wing had been burning for over six minutes![83] The ship could blow up at any moment.

Both Staff Sergeants Joseph K. Kocher, ball turret gunner, and Stanley Jarosynski, left waist gunner, jumped using the waist door.[84] Technical Sergeant Carl J. Brunswick slid out of the top turret and headed into the bomb bay to bail out. To his dismay, the doors had just been closed after the bombs left. He headed back onto the flight deck, ducked down into the bombardier-navigator section, and pulled himself through the nose hatch and free of the aircraft.[85] Hocker and Bullock followed.[86]

The Bullock aircraft crashed at approximately 1441 hours just northeast of Schweinfurt.

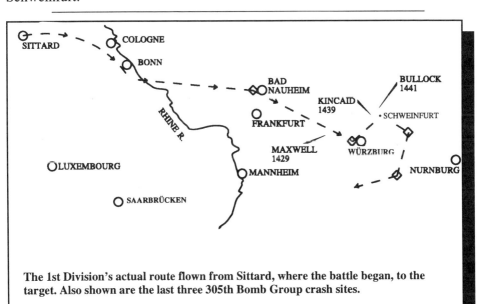

The 1st Division's actual route flown from Sittard, where the battle began, to the target. Also shown are the last three 305th Bomb Group crash sites.

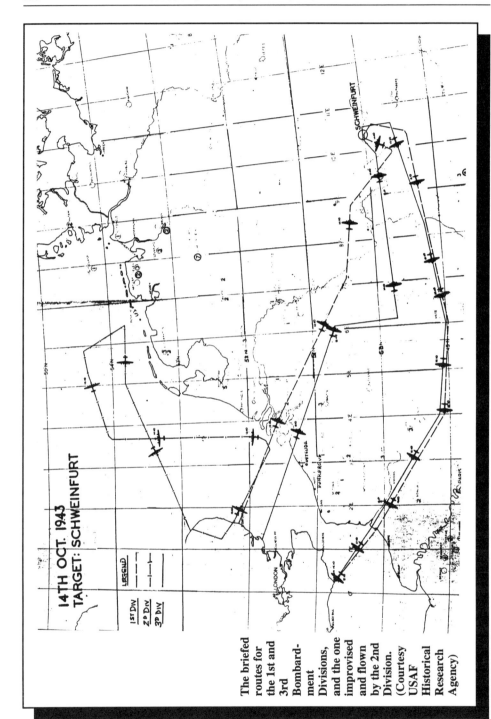

The briefed routes for the 1st and 3rd Bombardment Divisions, and the one improvised and flown by the 2nd Division. (Courtesy USAF Historical Research Agency)

Chapter 11

The Return Home

The diving turn off the target took the 1st Combat Wing and the two remaining aircraft of the 305th down to 21,000 feet.[1] Here they leveled off and began working their way back to England with a series of short doglegs, one to the southeast for 15 miles and the other to the south-southwest for 30 miles. After that, they would turn west towards home.[2]

The Farrell Crew

Just moments after the bombers completed the first turn to the southeast, a twin-engine Bf 110 pilot made a head-on pass at both the 305th aircraft. Several of his 20mm rounds tore into the nose section of the Farrell ship. One projectile exploded near the head of Lieutenant Frederick E. Helmick, the bombardier, and between himself and Lieutenant Max Guber. the navigator. Guber crumpled to the floor as he caught the full brunt of the blast in his face and chest. The bombardier's flak helmet prevented any damage to Helmick's head, but he also fell to the deck as numerous shell fragments ripped into his back. He lay stunned next to his bombsight, and for the moment he could not move.[3]

Another of the detonating rounds also ruptured a hydraulic line and damaged the oxygen system. Unknown to the crew, two of the four independent oxygen systems began to leak – the one the pilot drew from, and the one that fed air to the waist and tail sections.[4]

Meanwhile, Farrell sent Roy Burton, his copilot, down into the nose to find out the status of his navigator and bombardier. Burton crawled down through the hatch just into the nose section but went no further. The place was a mess. It was filled with maps and papers swirling about from the cold air which was streaming through the shattered Plexiglas nose. He saw both officers lying on the floor covered with blood from their wounds and hydraulic fluid which had sprayed from a broken line. Neither one of them was moving. Burton waited a few more moments looking again at the two prostrate flyers. Neither one seemed to be breathing. He quickly left and made his way back up into the cockpit. After strapping himself back into his seat, he plugged in his headset, and told Farrell it appeared Guber and Helmick were both dead.[5]

Shortly after Burton left the nose, Helmick realized he could move his arms and legs despite the pain of his wounds. He now made a supreme effort to reach Guber, who was lying on his left side next to the navigator's table.

The bombardier managed to crawl back to the lifeless form of his crew mate and rolled Guber over on his back. He could see his friend had been hit in the face and chest by the exploding shell. It appeared to him the navigator had lost both his eyes. In addition, the oxygen hose of his mask was shredded. Helmick glanced at Guber's oxygen flow indicator. It was not moving, and he realized the navigator was not receiving any air. On the floor by the table Helmick spotted another mask that Guber had somehow thought to bring along. Helmick was pleasantly surprised to find the oxygen mask, for an extra breathing device on board most B-17s was extremely unusual. Helmick now began to feel weak from his continual bleeding, but he kept moving. As best he could, the bombardier carefully removed the tattered mask from Guber's bloody face. He now adjusted the new mask to the face of the unconscious navigator. Helmick had to be especially careful, as he had to work the mask around Guber's protruding, damaged right eye. The bombardier again glanced over at the oxygen indicator, and this time it was moving. The new breathing device worked – and Guber was still alive.[6]

Despite his own painful wounds, Helmick continued to work as he now wrapped Guber's parachute around the unconscious flyer and made him as comfortable as possible. While tending to Guber, Fred Helmick noticed his own flying boots filling with blood.[7] He was slowly bleeding to death, and they still had to fly almost three and a half hours to get home.

The freezing air continued to pour into the nose section through the broken Plexiglas, but this proved to be beneficial to both wounded men. It assisted in greatly stanching the flow of blood from their open wounds.

The bombardier continued to stay busy. He found his headset and called up to Farrell in the cockpit. He informed the pilot he and Guber were still alive, and he had everything "under control" down in the nose.[8]

He did not mention the nose section was a "disaster area." Besides the damaged nose all the instruments, including the Norden bombsight, had been destroyed, and there was blood all over everything.

Next, the wounded bombardier crawled over to the navigator's gun position where he could peer out ahead from behind one section of undamaged Plexiglas. He looked just in time to see a pair of Bf 109s coming straight at his plane. They made their "classic attack" straight on at the nose firing as they came, then doing a half roll and falling away under the bomber's wing. Luckily, this time they both missed! They might return and the navigator's gun was empty, so Helmick began to splice two sections of ammunition belt that totaled about 30 rounds. While still working to get the ammo belts together, he noticed the same two fighters up ahead again, and they appeared to be getting in position for another head-on attack. In they came, only this time they did not fire, but just made their pass. Helmick figured they must be out of ammunition and were making practice runs or taking photos. He now finished the chore of forming one belt and was able to load the weapon. He now positioned the muzzle end of the machine gun on a spot ahead of the right wing where the Bf 109s had flashed by on their first two runs. He did not have to wait very long as the two fighters returned for a third pass. He looked straight ahead and here they came again. With his eyes on the fighters and not on the gun sight, he squeezed the trigger as the Bf 109s flew through the designated area where he had pointed the weapon. Both fighters immediately took numerous hits. One began to smoke, and parts began to fly off the other as the two German pilots were forced to bail out of their stricken aircraft.[9]

Helmick now turned his attention to taking over from the unconscious Guber. He began to plan the navigation needed to get back to base once they crossed into British airspace.[10]

After coming off the target, Milton, still leading the 1st Division, slowed his wing a bit to allow what was left of the 305th, and the remnants of Peaslee's two groups to get in as close as they could to him for supporting firepower. The bombers still had almost another two and a quarter hours to fly until they reached the French coast.[11]

Lieutenant Colonel Milton, had only lost three B-17s out of 33 in his wing to this point – one before the bomb run and two after. Colonel Peaslee's 40th Combat Wing had been decimated. Including the two 305th aircraft, his three groups now had a total of 20 aircraft remaining. The 40th Combat Wing had lost 29 B-17s – 27 before the target and two after! The 41st Combat Wing, still trailing the 1st and 40th wings had just bombed and began coming off the target. Many of the German planes now ripped into this rear box and began to pound its three groups, while the other two wings were receiving only sporadic fighter attacks. It was during this time most of the 41st's losses would occur. Prior to the bomb run, Lieutenant Colonel Rohr's wing had only lost one B-17. However, as it fought its way home, the 41st Combat Wing would be whittled to 36 aircraft from an original attacking force of 48.[12]

As the 1st Division continued to pull away from the target, the second prong of the bomber stream, the 3rd Division, which had been south of the 1st Division all day, took over and bombed Schweinfurt, from 1451 to 1457 hours. Their work finished, they also turned towards the west and now began to fight their way back to England.[13]

At 1510, twin-engine day fighters from the Wertheim area continued to attack both the 1st and 3rd Divisions together with the single-engine squadrons from Stuttgart and Munich areas. However, the fighter force was nowhere near the strength it had been early that afternoon.[14]

At approximately 1535, the two remaining B-17s from the 305th were still alongside the 1st Combat Wing. Milton, back up to 22,000 feet due to cirrus clouds, was now 24 miles southeast of Metz, France.[15]

This force had only completed about one-third of its trip home when the 305th had more problems. The Kane crew was holding its own against the enemy, but in Farrell's ship, there was a problem in the cockpit – the pilot could not breathe.[16] The two slowly leaking oxygen systems, which had been damaged earlier, began to take their toll. There was a mad scramble by the crew to keep Lieutenant Farrell from passing out. All the "walk around" bottles were quickly gathered, and as soon as one was empty, he was handed another.[17]

Some of the other crew members began to feel the effects of the hypoxia. Staff Sergeant Luther Bonones, at left waist, had been exchanging machine gun fire with two Fw 190s approaching from nine o'clock high when he suddenly stopped firing. He now began waving wildly at the enemy pilots. Fortunately. the fighters broke off the attack. In the tail section, Sergeant Niemiec came on the intercom talking about pretty neon lights. He was referring to 20mm cannon fire which was shredding the fabric from the rudder directly over his head.[18]

Down in the ball turret, Staff Sergeant James F. Higdon continued to blaze away, showing no ill effects from the lack of oxygen. He had his own auxiliary system. Fortunately for the two wounded officers in the nose of the ship, their oxygen system was also functioning normally.

German interception along the 1st Division's withdrawal route had not been planned as well as the tracking of the B-17s on the way into the target.[19] This was a godsend for the exhausted bomber crews, as many of them were down to their last few rounds of ammunition. On Kane's crew, several gunners reported they had run out of ammunition shortly after leaving the turret.[20]

German ground controllers now began to send squadrons of single-engine fighters to locations reporting stray bombers trying to make it home. Fortunately these "search and destroy" missions were not very successful.

The Return Home

At 1541, one group of 10-15 single-engine fighters swept across an area of Belgium but came up empty. Also at this time, many fighters were requesting permission to return to base. They, too, had become weary from the continuous air battle, and numbers of them were constantly running low on fuel and ammunition.

Another group of some 50 fighters was flying a patrol in the vicinity of Courtrai, Belgium but could not find anything to shoot at. There was much confusion among the German ground controllers as to where the bombers were located. At 1611 one searching fighter squadron landed in Belgium to refuel and reported it had not seen any B-17s along its flight route.[21]

The Germans were also flying similar patrols over France looking for the bomber stream and any possible stray B-17s. Unknown to the surviving crews of the 1st Division, they now received a blessing. The bombers had just missed another large battle with the enemy. At 1610, six German fighter squadrons had roared through French airspace, south of Rheims, without detecting any B-17s.[22] At 1611 hours, one minute later, the 1st Division passed just 14 miles south of Rheims, France. They were that close!

Milton now began a turn to a northwesterly heading of 302 degrees. This would take the 1st Division to the English "coast in."[23]

As the bombers turned on their final leg for home, the crews began looking for their "little friends." American P-47s were scheduled to rendezvous with the "heavies" in the vicinity of Rheims and escort them home.[24] In addition, British Spitfire fighters, were to meet and provide additional support for the bombers some 48 miles further ahead in the vicinity of Amiens, France.[25] However, friendly fighters were no where to be seen.

The Eighth Air Force weathermen had missed again. Their prognosis, for good weather over England for the returning bombers, was a "bust." The rapid climatic changes over the British Isles had made it impossible for the American and British fighters to get off the ground and reach France to complete their return escort mission with the B-17s.[26]

At 1617, another group of German fighters reported it was 18 miles southeast of Brussels and had no contact with the Americans. The fighters, who were combing the Belgian sector, were now recalled.[27]

The German fighter patrols flying through France reported some successes. German planes were reported attacking and destroying several stray bombers around 1645 hours.[28]

The 1st Division continued to receive a few attacks as it neared the French coast. However, most of the German fighters continued to look in the wrong places.

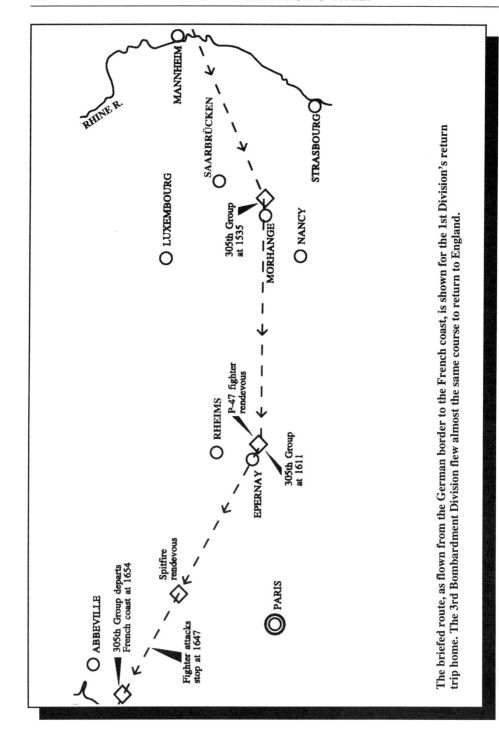

The briefed route, as flown from the German border to the French coast, is shown for the 1st Division's return trip home. The 3rd Bombardment Division flew almost the same course to return to England.

The 1st Division passed Amiens, and shortly thereafter at 1647, when the bombers were about 17 miles from the coast, the fighter attacks stopped.[29] The cessation of hostilities at this point in time was another blessing for the B-17s of the 1st Division. At this time, they passed about 19 miles southwest of Abbeville. France, which was home for one of Germany's most aggressive and successful fighter outfits – JG 26, the "Abbeville Kids." Today, for some reason, the "kids" did not show up.

Seven minutes later at 1654 hours, Lieutenant Colonel Milton and the 1st Division began to cross over the French coast for the 20 minute flight across the English Channel and home.[30]

Back in England numerous medical facilities had already been alerted to prepare to receive any wounded from the returning bombers when the aircraft landed at their various fields. The 49th Station Hospital, which was located 30 miles from Chelveston and served the Eighth Air Force as a primary medical facility, now made preparations for a great number of immediate surgical requirements. In addition they prepared for a large number of beds that could be needed.[31]

The 1st Division now began a gradual let down from 20,000 feet so as to arrive at the English coast at 11,000 feet.[32] As the bombers neared the "coast in", their crews could see another enemy awaiting them – poor flying weather.

The original forecast for the bombers' return to base had called for no clouds and visibility from four to six miles.[33] Nevertheless, the weather had changed and a new forecast was passed to the pilots. It was now reported a 1000 foot ceiling of stratus clouds existed and visibility was between one and two miles with haze.[34] However, depending on which airfield the pilot was trying to find, this "new" forecast was extremely optimistic, as in many areas the existing weather was much worse.

The word was passed. All ships would have to make individual instrument letdowns through the clouds, break into the clear, and then look for something that resembled home.

With the poor visibility, everyone had to be on the lookout for the other guys, who were also letting down and looking for their runways. As all the ships headed for their air bases, the small bit of sky under that low lid of stratus was going to be very crowded. These three elements of poor visibility, low ceilings, and crowded skies sometimes formed an equation that was a bomber crew's nightmare – a midair collision.

Due to the miserable existing weather conditions, a number of the pilots could not locate *any* bases upon which to land. As a result, three crews bailed out over England, while another two ships were crash landed.[35] In addition, two ships from Milton's 91st Bomb Group, and nine from Peaslee's 92nd Group had to settle for

305th Bomb Group medical ground personnel watch as their returning formations of B-17s approach the field. Ships firing flares had wounded aboard. (305th Bomb Group photo)

landings at other bases.[36]

After crossing the English coast, the remaining two ships from the 305th broke away from the 1st Combat Wing and headed for Chelveston via their prearranged letdown procedure. They completed their individual letdowns through the clouds without incident and turned towards their field as they broke out of the undercast. Since they were no longer flying together, each ship had to navigate its way to the landing runway. For the Kane crew this was no problem, but aboard the Farrell aircraft Max Guber, the ship's navigator, was still unconscious. However, Fred Helmick, the wounded bombardier, once again rose to the occasion. He managed, without navigational equipment or maps and with very poor visibility, to find some familiar check points and showed Farrell the way home.[37]

At Chelveston the ground crews, medics, and other personnel watched as Lieutenant Kane made his approach and landed at 1807 hours.[38] Two minutes later, at 1809, the Farrell aircraft touched down.[39] Due to the dismal weather, neither crew saw the other until the later debriefing.

Kane and Farrell parked their aircraft in their respective squadron hardstands, and shut down the engines. Ambulances immediately pulled up by the ships to evacuate any wounded.

Miraculously the group lead aircraft had no wounded aboard, with only Kane having singed hair and some superficial face bleeding caused by flying Plexiglas from the earlier cockpit explosion. Technical Sergeant Owen R. Hanson, the radio operator, and both Staff Sergeants B.T. Davis, ball turret gunner, and right waist gunner, G.M. Roe, had been extremely lucky. They had run out of ammunition shortly after leaving the target and so had no way of defending and fighting back.[40] Fortunately, at that time, the German fighters decided to pick on someone else.

At the Farrell aircraft, it was another story. Guber was still unconscious but breathing. Helmick was conscious but in a lot of pain. First aid was initially administered in the nose of the ship. As Helmick's flying boots were cut away, they were found to be filled with blood – he had almost bled to death.[41] Ever so gently the medics now eased the wounded officers through the narrow nose hatch, into the ambulances, and departed for the 49th Station Hospital.[42] The remaining members of the two crews gathered their gear together, and prepared to climb aboard the "two and a half ton" trucks that would take them to the mission debriefing building.

Farrell's radio operator, Technical Sergeant Roger J. Goddard, took one last look around before he left the aircraft. He could not believe his eyes. The radio compartment looked like a sieve – there were holes everywhere. He looked himself over carefully to make sure he had not been wounded. He had not.[43] Sergeant Smart, the right waist gunner, took a long circular walk around Farrell's plane and it appeared to him the wings and fuselage looked like large pieces of Swiss cheese.[44]

As the weary crews disembarked, the 305th commanding officer, Lieutenant Colonel McGehee, accompanied by his intelligence officer, pulled up in his jeep by Kane's aircraft. Only two 305th bombers had returned and landed, and McGehee was naturally puzzled as to the whereabouts of the others.[45]

He stopped Lieutenant Pellegrini, who had just slid out the nose hatch to the ground and was walking around the nose of the ship. They exchanged salutes, and McGehee asked, "Lieutenant, where are the rest of them?"

Pellegrini's tired reply was polite, but to the point, "Sir, there are no more ships. We are the only ones left!"[46]

The two exhausted crews then moved inside and, attended a very short debriefing.

Eight crew members of the returning Farrell crew standing beneath the damaged rudder of B-17F #230678. From right to left: Jayson C. Smart (with back to camera), right waist gunner; Thaddeus J. Niemiec, tail gunner; James F. Higdon, ball turret gunner; Ray A. Burton, copilot; Roger J. Goddard, radio operator; Luther Bonones, left waist gunner; Frederick B. Farrell, pilot; unknown person wearing dark jacket in background; Marvin D. Shaull, engineer and top turret gunner, wearing light colored overalls and ball cap. The two severly wounded members of the crew, navigator Max Guber and bombadier Frederick E. Helmick, had already been taken by ambulance to the 49th Station Hospital. (Photo courtesy of Frederick B. Farrell)

Meanwhile the many heartbroken ground crews, who were very close to the aircrews, and whose ships had not showed up as yet, continued to hang around their maintenance tents to park their respective aircraft if they arrived. They waited quite a while for more word and kept hoping there were some stragglers still on the way in, but no other ships ever returned. A short time later the official word was issued to the ground crews – the other 13, 305th B-17s, were gone and would not be coming home.[47]

BOOK II
AFTERMATH

Chapter 12

The Next Day

In the Morning

As the sun rose over the continent, enemy authorities across Holland, Belgium, Germany, and France were busily sifting through the wreckage of the many destroyed Eighth Air Force B-17s. They were looking for the remains and identity of the numerous dead American flyers who never made it through yesterday's air battle.

Some aircraft were still smoking and burning. In many cases there was not too much looking to be done. Many of the B-17s had burned leaving only scorched piles of metal and the charred bodies of crew members.[1] Here there was little to find.

A number of the aircraft had exploded on impact from the 6000 pound bomb load and the heavy load of fuel they were carrying. What was left of these people and planes was scattered across the surrounding landscape in tiny bits and pieces. Here there was nothing to find.[2]

A number of the dead Americans could not be identified due to mutilation caused by the burning and exploding planes. However, in most cases, the individual's "dog tags" were found on the remains of the deceased and used to provide positive identification. These two small identical metal plates, worn on a chain around the neck of military personnel, were the best method available to establish identity when death occurred in combat. One tag remained on the chain with the deceased, while the other was kept by appropriate authorities for official confirmation.

Whether identified or not, shallow graves were prepared for the dead airmen. Some received temporary wooden markers containing their names, while others were placed in unmarked spots in local cemeteries. In some cases, mutilation was so devastating, two or more crew members were inadvertently buried together as one.[3]

Most of the fliers who had bailed out and survived had been rounded up and were in the hands of the local or German military authorities. However, in the confusion of many planes and hundreds of American airmen coming down all over the terrain, the Germans were having a difficult time getting an accurate count. As a result, many who survived the bail out and "hit-the-silk" were still on the run in an attempt to avoid capture.[4]

With so many crews bailing out, the airmen of the five returning ships from the 306th, a sister group of the 305th, had described the scene as similar to a parachute invasion.[5]

Back in early morning England, a meeting of all wing and group commanders of the 1st Bombardment Division was scheduled for 1530 hours. At the 305th base at Chelveston, Lieutenant Joe Kane, who had been Major Normand's pilot on yesterday's mission was summoned to group headquarters. He entered the building and reported to the adjutant's office. A master sergeant, seated at a nearby desk, rose and handed Kane a piece of official looking paper.[6]

He then asked the pilot, "Sir, would you read and sign this citation recommending Major Normand for the Distinguished Service Cross for yesterday's mission? The Major has already been in and recommended you for the Silver Star!"[7]

Unknown to Lieutenant Colonel McGehee, the commanding officer of the 305th Bombardment Group, Major Normand had written this proposed citation for Major Normand.[8] He was recommending himself for one of the nation's highest combat decorations, one that is awarded for extraordinary heroism during operations against an armed enemy. It is second only to the Congressional Medal of Honor. However, a witness had to verify the facts and sign it.

Kane slowly began reading the contents of the paper and was astounded at what he saw. Normand had filled his proposal with self-seeking fabrications concerning his conduct as yesterday's group leader. To Kane, the recommendation reminded him of a publication he used to read as a teenager called, "G-8 and His Battle Aces."[9] He finished reading Normand's fable, and the sergeant held out a pen for the pilot's signature. Kane ignored the pen, and returned the unsigned proposed commendation to the noncommissioned officer.

"Sir," he asked, "Aren't you going to sign this?"

"No sergeant," replied Kane, "I sure as hell am not! And by the way, when you return it to Major Normand, tell him I said he can take this citation and stick it up his ass! As for him recommending me for a Silver Star, forget it . . . I already have one."[10]

With that, the conversation ended as the Lieutenant departed the headquarters.

The ship flown by the Farrell crew went through a thorough inspection. It had incurred an incredible amount of battle damage, but the bomber was repaired and flew again. It was later listed as missing in action on 25 February 1944.[11]

The lead B-17 flown by the Kane crew was also badly damaged and inspected. A partially unexploded 20mm shell was found imbedded above the instrument panel in the cockpit. The 305th ordnance officer reported the steel composing the shell was of a definitely inferior grade which prevented its complete detonation.[12] This aircraft was also restored and returned to service. Some two months later, on 21 December 1943, it did not return from a mission and was listed as missing in action.[13]

Over at the 49th Station Hospital at Diddington both Lieutenants Helmick and Guber, from the Farrell crew, were recovering from the wounds they had received immediately after the bomb run.

The pictures taken by the aerial camera aboard Farrell's ship of the 305th's bombing of Schweinfurt had been developed and analyzed. The lead bombardier, Lieutenant Joseph Pellegrini, had as his original aiming point a ball bearing factory.[14] By dropping on the leader of the 91st Group as Normand had ordered him to do, Pellegrini had abandoned his bombsight and flipped a toggle switch when he saw the bombs fall from the 91st.[15]

The photographic prints revealed the following: negative #1 -bombs away, and negatives #8 through #14 showed "approximately 18 bursts from our A/C [aircraft] in the northwestern-central residential area of the city of Schweinfurt." After struggling against overwhelming enemy fighter attacks for an hour and seven minutes to reach the target, the eighteen 1000 pound bombs of the remaining three 305th aircraft landed in a civilian neighborhood of the city of Schweinfurt.[16] Just as he had done all day long throughout the mission, Major Normand continued to be at the wrong place at the wrong time!

In the Afternoon

The Meeting[17]

At 1530 hours the wing and group commanders gathered in the conference room at the headquarters of the 1st Bombardment Division for the purpose of discussing yesterday's mission.

Brigadier General Robert B. Williams, the 1st Division commander, had called this meeting and was the host. Two very important guests were present, Brigadier General Frederick L. Anderson, commanding general of Eighth Bomber Command, and Brigadier General Orville A. Anderson, Eighth Bomber Command deputy commander for operations.

General Williams opened the proceedings with the following statement, "We are going into yesterday's mission in quite a bit of detail and are going to find out from what we learned by being there how to counteract that situation in the future. I want *no one to hold back*, and I want to hear all your ideas. We must build up countermeasures to *cut down such losses* in the future. . ."

This was indeed a great start to air out and resolve a lot of problem areas, especially those that prevented many of the bombers from flying a close, defensive formation. For despite the largest number of enemy fighters ever encountered by the bombers, most of the enemy deliberately avoided the B-17s flying as a tight unit. Instead they worked over every straggling bomber they could find – and they had an abundance of targets!

Colonel Marion, 1st Division Staff, began with a few opening remarks concerning the good coordination made with the 3rd Division concerning assembly over England. He also commented on the division lead changing hands as he stated, ". . . The 40th Combat Wing was to lead, but one of their groups did not get in their formation so the *Air Commander* instructed the 1st Combat Wing to take over the *lead*."

Throughout the afternoon, as the various commanders rose, one after another, to give their version of the Schweinfurt raid, any reference to mistakes that occurred or the poor judgment exercised by some air commanders was completely avoided. Much praise and backslapping took place, and half-truths, distortion of the facts, and inaccurate statements abounded. Many who spoke made sure they invoked the old military custom known as, "Cover your ass!"

Colonel William M. Gross, commander of the 1st Combat Wing, took the floor next and for good reason. He obviously wanted to protect Lieutenant Colonel Theodore Milton, who had assumed command of the 1st Division on his own volition while on the way to the target. Gross had not participated in the mission and of course was giving his version, to everyone present, of what he had been told by his own people:

". . . Two of the groups got together nicely. The other [the 381st] did not make the assembly until the point of departure [Orfordness]. A few minutes off the coast they were together and about mid-channel Col. Peaslee turned over the lead to my combat wing. The first decision to be made by the combat wing leader, Lt. Col. Milton, was whether or not, in taking over the lead, he also took over command. *He decided he did*, and went ahead and ran the mission from there on. I think it would be better for Lt. Col. Milton to take over from here on. In my opinion he did everything possible and did a fine job."

Colonel Gross' statement concerning the 381st Bomb Group was far from correct. That unit did not assemble at Orfordness with its wing, and "a few minutes off the coast they were [not] together." The 381st found Milton much later – some 30 minutes later. The 305th, led by Major Normand, arrived in the low position reserved for the 381st, at 1301, shortly before reaching the enemy coast. The 381st arrived after that, and had to take a position high on the right of Lieutenant Colonel Milton's high group, the 351st.

To hear Colonel Gross tell it, the 1st Combat Wing was intact when Colonel Peaslee assigned the lead to Lieutenant Colonel Milton. It was not! At that time, Milton had the same problem as Peaslee – he was short one group! However, Milton did not relay this information to Peaslee.

Gross' defense and support of Milton, his subordinate, who broke the cardinal military rule of usurping command, is unexplainable. Neither senior commander present, Anderson nor Williams, had any comments on this serious breach of discipline. Colonel Gross now turned the meeting over to Lieutenant Colonel Milton.

Milton began, and during this meeting the only referral he made to taking over the lead and command from Peaslee was, "... After we finally got our low group in, we were a little late starting the climb. Col. Peaslee called and said he was coming in behind to join us..."

Another half-truth, this time from Milton! Peaslee began moving his two groups of the 40th Combat Wing in behind Milton's two groups of the 1st Combat Wing at 1251 hours.[18] Ten minutes later, at 1301, Normand pulled up in the low group position of Milton's 1st Combat Wing which belonged to the 381st Group. Normand, now beside Milton, decided to stay there.[19] At this time, Milton's 381st Group had still not arrived on the scene!

Milton continued and gave a rather detailed summary of the mission, the fighting, and most of what occurred throughout the day. He failed to mention he assumed the lead from Peaslee while he, too, was missing one of his units, and gave no explanation as to why he took command away from Peaslee. He also omitted any reference to the off course overflight at the heavily flak defended areas of Cologne and Bonn. In fact, this unpleasant topic, like a number of others that should have been addressed by the senior officers present, was avoided by all of them.

The next item on the agenda was a discussion of enemy fighter tactics between General Williams, General Orville Anderson, and Lieutenant Colonel Milton. Milton went into great detail on the fighter tactics used against his three groups of the 1st Combat Wing. However, the Luftwaffe had very little success against this box as Milton only lost one ship from each of his three groups throughout the entire mission.

Milton concluded with the following statement, "... Although the enemy fighters were there, they didn't concentrate on us from the IP in to the target."

With that, Colonel Gross began the backslapping. Again, he had not been there, but he provided more lip service.

"Just one more thing. I would like to sound off my appreciation for the way Col. Peaslee brought his combat wing up and joined us. All my leaders greatly appreciated his support."

General Williams was now caught up in the turnabout of the meeting's purpose. He also praised Peaslee, whose units lost almost half of the entire attacking force, and said, "It looks like you cut down what might have been considerably more losses, Col. Peaslee."

Colonel Howard M. Turner, commander of the 40th Combat Wing, now had Colonel Peaslee give his version of the air battle.

Peaslee began, "As air commander I was riding in the lead ship of the 92nd Group as copilot . . . We picked up one group [306th] there [at the assembly splasher], but didn't see the other. We had some trouble with VHF and the other group [305th] couldn't get his transmissions through. As a result we started on course with two groups. At that time I could see the 1st Combat Wing, but I thought it was the 41st. Rather than lead the division with only two groups, I decided I would tie in behind another combat wing and go in the center of the division formation. I assigned the lead initially to the 41st Combat Wing for I thought they were the ones out in front [of the other two boxes]. However, it turned out to be the 1st Combat Wing, so I transferred the lead over to them. *I didn't relinquish command of the division, just the lead.*"

Generals Anderson and Williams had just heard it again! The 1st Division headed towards Schweinfurt with two air commanders. There were no comments from either one of them.

Now Colonel Peaslee picked up from Colonel Gross and continued with the backslapping as he said, "I would like to complement the leader [Milton], for he did a magnificent job of *leading* the show. . ."

Peaslee now continued with his version of the enemy fighter attacks against his wing going into great detail concerning the German use of aerial rockets against the bombers.

Colonel Peaslee concluded this portion of the his comments with the following statement, ". . . I also forgot to mention that one of our groups [the 305th] was flying low position with the 1st Combat Wing and remained there throughout the entire mission, until it had dwindled to two ships."

Peaslee went on, giving more graphic descriptions of the enemy attacks on his wing. The topic of German fighter tactics was discussed again by several senior officers. This time it was Colonel Peaslee answering the questions of both General Williams and General Orville Anderson.

Brigadier General Robert F. Travis, commanding general of the 41st Combat Wing, was next. He, just as Colonel Gross of the 1st Combat Wing, had not flown on the mission. He briefly discussed the 41st's participation in the raid, "... There was some trouble with the overcast, but the groups got together well ... I believe the losses [of the 41st] were held down by the good formation flying..."

It should be noted, the entire three groups of the 41st Combat Wing lost a total of 13 aircraft, which was the same number given up by the misled 305th Bomb Group.[20]

General Travis continued, "We saw contrails of our P-47s about fifteen miles off the enemy coast, and estimated them to be about 5,000 ft. above. We got no support, but we didn't need it at that time. I think they left about at the Dutch border. We had no enemy opposition until about one hour before the target ... Flak over the target was reported very light and they [his wing] were not under fighter attack from the I.P. to the target. We had an interval of about ten miles to begin with, which we later closed to about two miles. All of our losses were due to fighter attacks ... We had two ships from which the crews bailed out at sea, but they aren't back yet. There were mighty few holes through [clouds] which they could let down."

Unless the two most senior commanders present, Generals Frederick Anderson and Robert Williams, were completely in the dark concerning what happened on Mission 115 to Schweinfurt, neither one of them seemed concerned over what took place above the English Channel on the way to the target. This is where the big problems began, continued to worsen, and placed many bomber crews in jeopardy. These two generals seemed much more interested in the various tactics used by the Luftwaffe as it picked apart the straggling B-17 crews, many of whose misfortunes were caused by their own air commanders. It had been a bloody, costly battle, and the bombing had caused considerable damage to the target. However, despite the hundreds of German aircraft that swarmed over the 1st Division, many bombers and their crews were lost unnecessarily.

It must have been obvious to those present, the two units that performed the poorly executed maneuvers over the English channel, the 40th Combat Wing and the 305th Bomb Group, were the ones that were later crushed in the air battle. None of the senior commanders showed any interest in why many of the aircraft became strung out and were picked off one by one. Even Eighth Fighter Command had received word from their P-47 pilots, and had officially told Eighth Bomber

Command, ". . . the bomber boxes were reported to be considerably strung out."[21]

It was not the responsibility of the subordinate combat commanders present or the staff members in attendance to begin censuring one and other at this meeting and asking the embarrassing "hard questions." That was a command responsibility of the two senior officers present who had nothing to lose: General Robert Williams, who commanded the 1st Division, and General Frederick Anderson, who ran the entire bomber show! Either one could have picked up on the mistakes made, and offered guidance and constructive criticism for the others to follow in the future.

Instead, General Williams never followed through on his opening remarks concerning building counter-measures to cut down heavy bomber losses in the future. These good intentions of his were now derailed, and they never would get back on track. Williams, the host and moderator, simply let the meeting get away from him. General Anderson, whose command had suffered its greatest defeat to date, had now lost air superiority to the enemy![22] Despite this, throughout the gathering, he continued to sit and listen. Since Fred Anderson did not question anything, neither did General Williams, his subordinate. If there were combat leaders and staffers present who were looking for guidance from above, they all came away empty. The 1st Division critique had completely lost its objectivity.

The meeting now turned to a discussion on radio procedures to include blocking the weather channel with too much chatter, and poor radio discipline. Several comments were made concerning the dropping of auxiliary bomb bay fuel tanks.

General Williams now spoke, "Is the leader of the 305th here? Do you have anything you would like to add?"

Major Normand stood, and addressed those present, "I would like to mention one thing. We circled Thurleigh at 12,000 ft. in a very shallow turn and I saw at one time four groups. I kept calling on VHF for identification signals, but never did get any response. We were already a bit late [nine minutes], so I told the navigator to give me a course for Orfordness, where we arrived one and a half minutes late. Two groups were there and I thought they were the rest of our Combat Wing, but it turned out to be the 1st Combat Wing. Instead of milling around, I thought I would just tack on as low group, which position I was to fly in my own combat wing, and go on."

Major Charles G.Y. Normand's remarks concerning how he arrived at Orfordness and when he met and joined the 1st Combat Wing were outright lies. However, no one challenged him!

He did not proceed directly from Thurleigh to Orfordness as he so stated, but instead headed west for Daventry, cut that short, proceeded to Spalding, and then went on to Orfordness.

Earlier, on this same day, 15 October 1943, Normand's navigator, Lieutenant Jack J. Edwards submitted a "Description of the Navigator's Problem on the mission of 14 October 1943" in writing to the Commanding Officer of the 305th, Lieutenant Colonel Thomas K. McGehee. It read, "At the time of wing assembly, three or four groups were visible flying around Thurleigh, but no flares were fired. While waiting for flares to identify our wing, we ran a bit short on time, and proceeded on to Daventry where we expected to join the wing. Upon *nearing* Daventry, no groups were seen so we proceeded to Spalding, our next point, where we arrived prior to the briefed time by about six minutes. We circled over and as only a complete wing was visible, we were forced to proceed on to Orfordness. . ."[23]

Normand continued to mislead those present with his untrue version of what took place upon arriving at Orfordness. The 305th leader of yesterday's mission was telling everyone present he joined Milton's 1st Combat Wing at Orfordness. In reality, he had taken the 305th Group, all by itself, far out over the English Channel thinking the 1st Division was in front of him.[24] Neither he nor anyone else in his three squadrons saw the two groups he claimed were there when he arrived. He identified two nonexistent groups as part of the 1st Combat Wing. At the time he stated he joined them, they were many miles north of him trailing his leader, Colonel Peaslee. He did not join the 1st Combat Wing until well after Peaslee had turned the division lead over to Milton's outfit. This was some 29 minutes later, 1301 hours, over the channel, and actually three minutes out from the enemy coast.[25]

Lieutenant Edward's report substantiates this! It went on, ". . . We left the English coast at Orfordness two minutes late and without having joined our wing. About 30 miles short of the enemy coast, we sighted two groups following us, and [we] started an "S" [turn which in this case turned out to be a 360 degree turn] to fall in with them under the assumption they were our wing. After joining them [we] discovered they were the 351st and 381st groups [he meant the 351st and the 91st groups of the 1st CBW]. We joined them as low group, and proceeded on the briefed route to the target."[26]

Normand completed his discussion as he briefly mentioned the German fighter tactics he had seen.

Both Peaslee and Milton had to be aware Major Normand was lying but both remained quiet. When Peaslee led the 1st Division to Orfordness, Milton was well behind Peaslee in the #2 position with his 1st Combat Wing. Peaslee even made a 360 degree turn, at the coast, to enable the other two wings of the division to close up the formation. If Normand had seen Milton at that time, he would have also seen Peaslee and the remainder of the 1st Division.

At this meeting Major Normand was not queried as to why he never found his wing, and why he suffered such astronomical losses. Normand dropped 13 out of 15 bombers to the enemy, while being a fourth group of the leading 1st Combat Wing which only lost 3 ships all day![27]

Another discussion now began between General Orville Anderson, Colonel Peaslee, Colonel Marion, and General Williams on the subject of the German fighters firing aerial rockets at the bombers.

Colonel Turner again spoke up, "Major Normand has one more point."

This time Major Normand's comments involved his bombardier, Lieutenant Joe Pellegrini, who had begged Normand to let him make a separate bomb run with what remained of the 305th Group. Just as Edwards had done, Lieutenant Pellegrini had also signed and filed an official statement earlier in the day with the commanding officer of the 305th.

Pellegrini's report stated, ". . . When the group turned on the bomb run, the bombardier called the pilot [Normand] and instructed him to turn on the run. But the AFCE had been shot up, and he [Normand] was not sure the PDI was working. He also advised that due to the small force left, four aircraft, it would be better to attach ourselves to the group ahead. During this time we were headed for the aiming point upon which the bombardier had already set his cross hairs and was synchronizing on. At this time the fighters concentrated in ever increasing strength upon our formation compelling us to attach our remaining aircraft to the lead group. . ."[28]

To all those present at the 1st Division get-together, Normand now gave a version completely different from Pellegrini's written report of what took place on the bomb run.

Normand said, "When we got over the target, I only had three ships left in the formation. The bombardier called me and requested that I drop back so we could make a run of our own. However, when *he* found out we only had three ships, *he changed his mind. . .*"

So according to Normand, his bombardier had made the decision not to bomb as a group! Again, Anderson and Williams listened, but said nothing.

Colonel Turner commanding officer of the 40th Combat Wing, which had been clobbered, was present throughout the entire meeting. He had nothing to say to those present concerning how his wing had conducted itself.

After a short discussion on the malfunctioning of bomb bay tanks, General Williams introduced the leader of Eighth Bomber Command. General Frederick L. Anderson rose, and finally prepared to speak. However, what he had to say had nothing to do with the yesterday's mission critique that had just ended at 1st Division

Headquarters.

He was about to tell his commanders, though not directly, he and his seniors had finally seen that hand writing on the wall. The omen of one's unpleasant fate could no longer be ignored, and the Eighth Bomber Command folly was about to end!

"I have spent many afternoons in this room, but haven't been up here for a long while, for which I am sincerely sorry. I've tried to get here several times, but just couldn't make it. Yesterday the 1st Division took a very heavy blow . . . The 3rd Division took a heavy blow last week at Munster. This month we have lost in Bomber Command 150 heavy bombers.[29] It sounds like a lot, but let's look at the other side of the question. The Hun has lost well over 600 fighters of various types, besides his factories at Marienburg and the component parts factory at Anklam, which I consider a most important blow. We did a tremendous amount of damage at Gdynia and Bremen. He is still squealing about Munster. But the best attack of all was the one yesterday . . . Our losses were pretty heavy. Temporarily we are going to slacken up, but that doesn't mean, and I don't want anyone to get that idea, that he [German fighter force] has hit us a blow that we can't recover from. I am going to slacken up because we have hit his fighter industry so heavily that we can afford to wait until attrition begins to catch up with him. *We can afford to come up only when we have our own fighters with us* . . . A newspaper reporter called this morning and commented on the heavy losses, and asked if we were going to stop. We are going back. You will be built up with crews and planes just as quickly as possible. You know, and I know, that all those men lost are something which we just can't forget."

Brigadier General Frederick L. Anderson now made a verbal blunder. It was reminiscent of yesterday's message read at all bomber group briefings before the crews left for Schweinfurt.

At these meetings, he had sent the following news for his crews to think about, ". . . Your friends and comrades that have been lost and that will be lost today are depending on you . . ."

Anderson now delivered some callous words for his 1st Division commanders to take back to their bomber crews as he stated, ". . . I want you to go back to your crews and tell them just how worthwhile these losses are . . ."

He continued, ". . . The 305th ended up with two planes, the 306th with five, and last week the 100th Group from the 3rd Division ended up with one, – *but the bombs were on the target* . . . Gen. Eaker [Eighth Air Force Commander] would have liked to be here himself today, but he is out trying to get us the long-range fighters we need . . ."

General Williams now concluded the meeting with several closing remarks.

Since it was apparent there were no dues required to be paid in this club, it was now back to business as usual. The meeting adjourned at approximately 1700 hours.

Some days later, final photo interpretations of the bombing result estimates were completed. The principal targets attacked were the Kugelfischer Works and Vereingte Kugellager Fabriken Works I and II. It was estimated these three plants contributed 50% of the total output of ball bearings available to Germany. A careful study of reconnaissance photographs indicated the attack resulted in a loss of 75% of the productive capacity of the ball bearing industry at Schweinfurt.[30]

Later, this loss estimate was shown to be extremely over-optimistic. The raid had caused a machine damage of just 10%.[31]

Chapter 13

The First 12 Minutes of Battle

In order to fully understand the devastation that occurred to the 305th Bomb Group, it is necessary to reconstruct the time frame during the destruction of the planes and crews of the high and low squadrons. By his inept leadership, Major Normand set the stage for the Luftwaffe to clobber his scattered group. This air commander's mistakes greatly contributed to the loss of the above two squadrons and caused many of the crew members in these units to suffer almost every conceivable misfortune from the overwhelming German fighter attacks.

For the 305th, the air battle started at 1333 hours. Seven minutes after the fighting commenced, the 305th began to come apart. In just five minutes, from 1340 until 1345, four of this group's B-17s crashed west of the German border. Three went down in Holland and one in Belgium.

The Murdock Crew

Ironically the Murdock outfit, of the low squadron, was the last one off the ground at Chelveston in the morning, and it was the first one, from the 305th, to go down that afternoon![1] At 1340 a Dutch farmer watched as the Murdock aircraft crashed and then exploded twice spreading fire over a large area near Limmel, Holland not far from Maastricht and just east of the Belgian border.[2]

The crew members that were able, hit the silk between 18,000-19,000 feet. Sergeant Lester J. Levy, the right waist gunner, had been fitted with a new harness for today's mission, and there was too much slack in the crotch where the leg straps circled the inside of his legs. When his chute popped, the loose straps between his legs snapped inward and snared his testicles. Despite his earlier problems in the waist section with the loose shoulder harness tacking, then being unable to force open the waist door trying to exit the aircraft, and now his present agony caused by the improperly adjusted leg straps, Levy had a special reason to be thankful he had made it this far. Prior to take off from Chelveston, Sergeant Levy had been assigned to fly the left waist position. However, the other waist gunner, Staff Sergeant Tony E.

Dienes, asked Levy to move over to right waist as he was more comfortable on the left side of the aircraft. Levy obliged, and the wounded Tony Dienes never make it out of the plane. When Lester Levy hit the ground, he just laid there unable to get up due to his recent injury. He was captured immediately and spent several weeks in the hospital recovering from the leg strap incident.[3]

As copilot Lieutenant Edwin L. Smith floated downwards, he counted some 12 parachutes in the air and watched helplessly as a Bf 109 came directly at him. The fighter passed within 25 yards of him but the enemy pilot did not fire.[4]

Smith, Staff Sergeant William B. Menzies, the 17 year old tail gunner, and Sergeant John W. Lloyd, from the ball turret, all landed safely. They were captured within 15 minutes of each other in the vicinity of Maastrich, Holland.[5]

Technical Sergeant Thelma B. Wiggins, the radio operator, had been wounded in the right heel by shell fragments just prior to bailing out through the waist exit. He landed right beside a canal nearly 16 miles southwest of his crew mates at a hamlet called Tongeren which was well inside the Belgium border. He did not realize it at the time, but landing this far into Belgium was just what he needed. He now went into hiding with some new found friends.[6]

Later that day, Technical Sergeant Russell Kiggens, the engineer, and Tony Dienes, who had swapped places with Levy, were found dead in the destroyed aircraft.[7] Kiggens was last observed stepping down from the top turret and seemed okay, but for some reason he never made it out of the ship.[8] The severely wounded and unconscious Dienes was last seen lying on the floor by the left waist and had no chance to jump.[9]

Unfortunately for Lieutenant John C. Manahan, the navigator, his parachute failed to open, and he fell on the dike of a canal. He was discovered lying with his unopened chute later that day by two civilian workers.[10] Later as Bill Menzies was waiting to be interrogated, a German soldier showed him Lieutenant Manahan's torn parachute, and a pair of coveralls with the officer's name tag on it.[11]

On 22 August 1979, William B. Menzies, who had been Murdock's tail gunner, wrote a letter to a Mr. Jausen in the Netherlands requesting any information Jausen could provide concerning what happened to the Murdock crew and its ship on 14 October 1943. The letter was passed to Peter H. Luijten, a World War II Aviation Historian. On 22 October 1979, Luijten provided an informal report to Menzies. One of the pieces of information in the correspondence suggested Manahan's parachute was shot through by an Fw 190, but this was never confirmed.[12]

Staff Sergeant John E. Miller, the togglier, also did not survive, but his cause of death was never determined. His remains were later found after the war in a temporary grave and positively identified by the Army Graves Registration Service.[13] It was rumored, but never confirmed, that Miller, after bailing out, had joined the "Dutch Underground" and was active with them until his death. Another report received by Menzies in 1979 from a Dutch historical group stated Sergeant Miller died on 26 March 1945 – some 17 months after being shot down.[14] This date is also unconfirmed.

Lieutenant Smith spent the remainder of the war in Stalag Luft 3 and Stalag 7B, while Sergeants Levy, Lloyd, and Menzies were imprisoned in Stalag 17B at Krems, Austria.[15]

The pilot, Lieutenant Douglas L. Murdock, was not accounted for until well after the end of the war. After an official investigation surrounding his disappearance, which lasted over five years, his official status was finally determined on 31 January 1949. It was concluded Murdock, if not already dead as a result of enemy gunfire, died in the crash of his plane near Maastrich, Holland, on 14 October 1943.[16]

After the crash of Murdock's ship at Limmel, records at City Hall, Maastrict indicated the remains of three of the Murdock crew members were buried in the Maastricht cemetery. Of the three, one was unknown while the other two airmen were identified as Russell Kiggens, the engineer, and Tony E. Dienes, the waist gunner. Murdock was not reported as deceased by the German government and was therefore listed as "missing in action."[17]

Later, his remains were positively identified along with those of Kiggens thereby ending the uncertainty of his disappearance.[18]

The McDarby Crew

After Lieutenant Dennis J. McDarby gave the order to bail out, only five crew members were able to safely depart the ship. They all landed in the vicinity of Eygelshoven, Holland.

The McDarby aircraft came down in two pieces at 1345 hours.[19] The fuselage settled just inside the Dutch border at Eygelshoven near a small pond used for rowing, while the front end of the bomber fell just across the border into Germany near the village of Finkenrath.[20] This cockpit and wing area, still carrying the bomb load, blew up on impact leaving a large crater 13 feet deep and 35 feet in diameter.[21]

On the way down, pilot McDarby could see parachutes all over the sky. He now became aware he had superficial wounds around his face and neck from fragments of Plexiglas from his shattered pilot's window. He also noticed he had lost a boot.

When he landed on the Dutch side of the border, he hit the ground hard and severely sprained the ankle that was bootless.[22] While lying on the ground he was met by a Dutch coal dealer, who happened by with his horse drawn cart filled with bags of this fuel. He offered to hide the pilot under the coal sacks, but McDarby declined replying he wanted to stay with his crew.[23] He was then given first aid by some Hollanders in the dressing room of coal mine, "Julia."[24] The people tending his injury asked him if the invasion had come, for they had never seen so many parachutes in the air at one time.[25]

The engineer, Sergeant Arthur E. Linrud, came down just inside the German border and landed in a meadow alongside an asphalt road. German civilians and soldiers had gathered on the road watching him descend.[26] Unknown to him, as he drifted downward towards the unfriendly crowd, a German sailor, home on leave, fired several shots at the American with his rifle. Fortunately, the seaman was a poor shot, and Linrud landed safely.[27] He was immediately captured as were Staff Sergeants Dominic C. Lepore, the tail gunner, Hosea F. Crawford, the radio man, and Benjamin F. Roberts, the ball turret gunner.[28]

Just after his capture Sergeant Lepore, who had been wounded earlier in the head by a 20mm shell, had his wounds treated by a local doctor.[29] The radio operator, Sergeant Crawford, landed on a barbed wire fence, and was also treated by the locals for numerous cuts and scratches.[30]

After the forward portion of the plane crashed and exploded just inside the German border, Lieutenant Donald P. Breeden, the copilot, could not be located and was listed as "missing in action."[31] Both Lieutenant Harvey A. Manley, the bombardier, and Lieutenant William J. Martin, the navigator, were last seen getting ready to bail out. They both had told Arthur Linrud, the engineer, they would follow him out the nose hatch.[32] For some reason they never cleared the aircraft. The remains of Lieutenants Manley and Martin were recovered from this same wreckage and buried.[33]

Sergeant Leonard R. Henlin, and Sergeant Robert G. Wells were killed, at their positions, during the air battle.[34] The remains of both these waist gunners were found at the wreckage site of the fuselage and tail section which had fallen across the border just inside Holland.[35]

Later that day the five surviving flyers were taken to the Municipal Hall of Eygelshoven where Lieutenant Smith of the Murdock crew joined them. They were then loaded into a small, open truck by their captors as about 150 Dutch spectators gathered. Despite the numerous German soldiers present, the onlookers smiled and waved at the Americans and gave them an ovation as the truck pulled away. The GIs responded by waving back!³⁶

Lieutenant McDarby was hospitalized late that evening in the officer's section of a nearby large warehouse used by the Germans for a hospital. The next morning he was awakened by the nurses as they pulled back the curtains surrounding his bed. He found himself in one huge ward filled with hundreds of wounded German soldiers from the North African campaign.³⁷

Lieutenant McDarby spent the rest of the war in Stalag Luft 3 while Sergeants Linrud, Crawford, Roberts, and Lepore were incarcerated at Stalag 17B.³⁸

Lieutenant Donald P. Breeden, the copilot, was never found. On 20-21 December 1948 and 4-5 January 1949, an investigation was conducted into the disappearance of this officer. Two German civilians, who had been at the scene of the crash, stated a third civilian, who was present, had placed the indefinable, torn remains of what he believed to have been two deceased persons into separate wooden coffins. The two Germans testified the remains found lying about the crash site comprised more than two bodies. On 11 April 1951, a board of review recommended the remains of the decedent be determined nonrecoverable. Though not proven, Lieutenant Breeden may have inadvertently been put to rest with his two fellow officers, Lieutenants Martin and Manley.³⁹

The Eakle Crew

With two engines out, another one smoking furiously, and raw gasoline trailing from the left wing, Lieutenant Gerald B. Eakle rang the alarm bell for the crew to abandon ship – there was no intercom!⁴⁰

The first four crew members to leave the aircraft left by the waist exit. They included the two waist gunners, Staff Sergeants Alfredo A. Spadafora and Lloyd F. Knapp, Jr., together with the radio operator, Technical Sergeant Donald H. Norris, and ball turret gunner, Staff Sergeant Robert L. Sanchez. As they descended, all of them were buzzed by German fighters, but they were not fired upon.⁴¹

Sergeant Spadafora had bailed out at 10,000 feet, and as the gunner drifted downward in his chute he looked all around. On the ground, as far as he could see, were fires burning from downed aircraft – some enemy fighters, but mostly large, black fires from destroyed B-17s. He also noticed he was missing his right boot. Spadafora came down on the roof of a home located in a small village. Just before hitting the housetop, he pulled his knees up to his chin and used his legs as a buffer to prevent any serious injuries. After landing on the rooftop, he and his parachute came to an abrupt stop for just a few moments. Then ever so slowly he and his parachute began a long slide off the roof. He ended up on the ground completely entangled in his chute. He was captured by the authorities immediately. Several elderly German men and women, who had been watching the descent, came forward to help him. There were tears in their eyes as they pulled the chute away from him and assisted him to his feet. He had several small cuts and abrasions from his collision with the roof, and he was carrying several shell fragments in his left arm from the air battle. Later that afternoon, Spadafora joined other crew members who had been captured including Donald Norris, the radio operator, Robert Sanchez, the ball turret gunner, and the right waist gunner, Lloyd Knapp, Jr.[42]

Staff Sergeant Herman E. Molen, the togglier, who had helped the wounded navigator, Lieutenant Charles B. Jones, leave through the open bomb bay, immediately followed him out at between 1000-1500 feet. Molen landed in a rutabaga field just inside Germany and was captured at once. Several minutes after being taken into custody by the authorities, Molen watched as Flight Officer Verl D. Fisher, pilot of the Fisher crew, landed just across the road from him. Later that day Molen was reunited with Jones and Technical Sergeant Barden G. Smith, the engineer.[43]

Lieutenant Gerald B. Eakle, the pilot, who had been trying to persuade his tail gunner, Wilson, to jump, followed the gunner out the nose hatch and pulled his rip cord as soon as he cleared the burning plane. He was about 1000 feet above the ground and had just a short time before he landed in a plowed field. Just before he hit the ground, Eakle saw one chute in the distance that was not fully deployed disappear below the tree line.[44] It was Staff Sergeant Lloyd C. Wilson, who did not want to jump, procrastinating again! He delayed pulling the parachute rip cord, and when he finally did activate his chute, it popped just as he hit the trees.[45]

The Eakle aircraft fell in Belgium, near the town of Eisden, at 1345 hours.[46] It crashed in the same field where the pilot had just landed – some 700 yards from him. The ship continued to burn and eventually blew up.[47]

After landing Eakle discovered he had received superficial wounds from cannon fire in his left leg sometime during the air battle. He was captured some three hours later, and was brought to a German interrogation center together with gunners Lester Levy and John Lloyd of the Murdock crew.[48] It is not certain if the copilot, Lieutenant Walter H. Boggs, tried to crash land the ship or just could not get out of the cockpit in time. He was last seen trying to leave the cockpit behind Wilson and Eakle. He apparently was not wearing his seat safety belt and shoulder harness when the B-17 crashed, for he was thrown clear of the aircraft and died instantly. The remains of Lieutenant Boggs were recovered and identified.[49]

Everyone from the Eakle crew was captured by the end of the day except Wilson, the tail gunner. German soldiers continued to search for him, but he was nowhere to be found.

Lieutenant Jones recovered from his wounds, and he and Lieutenant Eakle were imprisoned until the end of the war in the officers' prison camp known as Stalag Luft 3.[50] The noncommissioned officers of the crew, less Sergeant Wilson, were sent to Stalag 17B for the duration of the war.[51] The Germans never found Wilson![52]

The Willis Crew

When Lieutenant Charles W. Willis, Jr. ordered his crew to leave their disabled ship, he was flying directly over the Maas river. People on the ground observed that six chutes came out of the plane. Several of the crew members landed on the west bank, while the others fell on the east side of the river. The pilot, flying very low and trying to land the aircraft while under heavy fighter attack, struck a tree! The B-17 crashed and exploded on impact.[53]

The crash occurred at 1345 hours between the small town of Horn and the larger city of Roermond, Holland just a few miles west of the German border.[54] The crash site was some 10 miles north of Sittard where the air battle had begun just 12 minutes earlier.

The remains of Lieutenant Willis and his copilot, Lieutenant John J. Emperor, together with Staff Sergeant Nicholas Stanchak, the tail gunner, were found in the aircraft wreckage.[55]

Staff Sergeant John L. Gudiatis, the right waist gunner, recovered consciousness just before hitting the ground in an opened chute. He was amazed he was still alive as he could not remember leaving the aircraft or opening his parachute. Since he and Staff Sergeant Edward J. Sedinger were the only ones alive in the waist section when Gudiatis fainted from loss of blood, apparently the gunner who would not jump helped his buddy who could not jump. Sedinger had often remarked to Gudiatis, "If ever I have to jump, you're going to have to push me out!"[56]

As Gudiatis lay on his back on the ground, he looked up and watched as two Fw 190s kept circling the area above him. Soon some German civilians and a doctor arrived on the scene. Since the wounded American was wearing a medal around his neck, another German arrived with a Catholic priest. After receiving first aid, the badly wounded flyer was then taken to a local emergency center by ambulance. Two days later he moved by train to a hospital in Amsterdam where he had surgery performed for the first time. He later spent five months in three different hospitals before being discharged. He was then transferred to a prison camp in Lithuania, and he never again saw any of his crew mates.[57]

After he landed, and was captured by the Germans, Technical Sergeant Floyd J. Karns, the engineer, was taken to an area near the crash site and identified the body of Sergeant Sedinger, the left waist gunner, who refused the order to jump. It appeared Sedinger, who had delayed jumping, either bailed out too late or was thrown from the aircraft on impact.[58]

The other surviving members of the crew were captured within a short period of time. The two officers, Lieutenants Christian W. Cramer, the navigator, and Willard E. Dixon, the bombardier, and the three remaining noncommissioned officers, Staff Sergeants Bernard E. Snow, the ball turret gunner, Alan B. Citron, the radio operator, and Technical Sergeant Karns spent the rest of the war in German prison camps.[59]

While a prisoner, Citron incensed his captors by openly writing on prison stationery, "The German guards are bastards"! He also wrote they mistreated the prisoners, and they should be castrated for their evil deeds. For his defiance, he was placed under arrest for crimes against the Third Reich, and was scheduled to go on trial. Fortunately for Alan Citron, the camp had to be evacuated, and the trial was never held.[60]

Chapter 14

The Next Four Minutes

T he battle continued uninterrupted and in the four minutes, from 1346 until 1350 hours, three additional 305th aircraft fell out of squadron formation with major battle damage. This time all of them were over Germany. In the short span of 10 minutes, the high and low squadrons had dropped seven ships. Up ahead, Major Normand's lead squadron was closely following Lieutenant Colonel Milton's 1st Combat Wing and the 305th leader had still not lost a ship from his squadron!

The Holt Crew

The noncommissioned officers of this crew had no notification it was time to bail out. The four officers in the front of the aircraft were either dead or severely wounded from the head-on fighter attacks.[1]

The only apparent survivor in the cockpit area was the flight engineer, up in the top turret, Staff Sergeant Charles E. Blackwell. He was now surrounded by a raging cockpit fire from the burning oxygen system. He somehow managed to locate his parachute and succeeded in getting down into the nose section which was also completely ablaze. Blackwell went out the nose hatch and was now free of the terrific heat and flame of the stricken airplane. His oxygen mask had been completely burned away from his face, but fortunately his parachute was not damaged from the inferno. It opened about 16,000 feet. Blackwell's tattered and charred flying gloves were lost when his chute opened. However, by wearing them, he had saved his hands from serious burn injuries as he groped his way free of the aircraft.[2]

The two waist gunners, Staff Sergeants Floyd A. Lenning and Frank W. Rollow, together with the ball turret gunner, Sergeant William C. Frierson, managed to get out of the falling tail section of their ship after it separated from the fuselage. However, they only had a few, brief moments before they hit the ground in their opened chutes. They had fallen clear of the separated, aft portion of the B-17 while 400-500 feet above the terrain.[3]

Just prior to hitting the ground, Rollow glanced up quickly and could only see Lenning and Frierson above him. He wondered about Bowman and Letanosky, who had also been tangled up and pinned down with the other three gunners in the falling tail section.[4]

Sergeant Blackwell landed shortly after the others. On the way down he was buzzed by an Fw 190, but the pilot did not fire at him. He also watched in disbelief as three more B-17s from other units of the 1st Division fell to the enemy – one blew up, and two of them spun in. He was able to count just three parachutes that made it out of the trio of bombers. He drifted down and into a shallow pool of slush near the entrance to a coal mine. As he stood and began to remove his parachute and harness, he realized he had been wounded in his right leg by 20mm cannon fire. He was captured several minutes after landing and taken to Dusseldorf for interrogation. It was here, on the following day, a local doctor removed his right boot, which was filled with blood, and bandaged the leg wound. He had burns on his face and hands which the doctor also treated. It would be another six months before the Germans would take the time to give him proper treatment for his infected leg wound![5]

At 1350, the front portion of the bomber crashed close to the hamlet of Immendorf, Germany. It burned, and the bombs later exploded from the fire. The tumbling tail section landed shortly thereafter not far from the front part of the burning airplane.[6]

The remains of the four officers: Lieutenants Robert W. Holt, the pilot, Aubrey C. Young, Jr., the copilot, and the navigator and bombardier, Lieutenants Edward O. Ball, and Bryce Barrett were later recovered from the wreckage.[7]

Technical Sergeant Doris O. Bowman, the radio operator and Staff Sergeant Mike S. Letanosky, the tail gunner, also did not make it. Two Germans, who worked for the mayor's office in Immendorf, found the bodies of the radio operator and tail gunner near the wreckage of the tail section still wearing their unopened parachutes![8]

Sergeants Rollow and Frierson were captured immediately, as was Sergeant Lenning. All had landed in nearby fields. However, Floyd Lenning had the misfortune to fall into the hands of the Hitler Youth, an adolescent group of 15-18 year old German boys used to exploit Hitlerism.[9] He was severely beaten by his captors, suffering a fractured jaw and several cracked ribs, before the military authorities arrived and took charge.[10]

The four surviving members of this original ten man crew, Sergeants Blackwell, Frierson, Rollow, and Lenning, were imprisoned together in Stalag 17B until the war ended.[11]

The Dienhart Crew

At 1349, several minutes after ordering his crew to bail out, Lieutenant Dienhart's B-17 went into a steep dive. One engine was feathered, most of the Plexiglas nose of the bomber was missing, and raw gasoline was trailing from the right wing.[12]

The Fisher Crew

Shortly after the Fisher B-17 slammed into the ground at 1350 near Waldenrath, Germany, the remains of the togglier, Staff Sergeant Donald L. Hissom, together with those of Staff Sergeant Harold Insdorf, ball turret gunner, and Staff Sergeant Thomas E. Therrien, right waist gunner were all found near the wreckage of the plane. Both Insdorf and Therrien had been killed during the air battle but Hissom's cause of death was never determined. He was last seen by the copilot standing in the cockpit with the navigator, Lieutenant Carl A. Booth, Jr. Both Hissom and Booth were okay and getting ready to bail out.[13]

The body of Lieutenant Booth was never recovered. Missing Aircrew Report #916 indicates Harold Insdorf was buried with three other unidentified airmen in unmarked graves. It is believed, but never confirmed, Carl Booth was initially interred with the above three gunners by a German burial party who believed they were putting three not four Americans to rest.[14]

The copilot, Lieutenant Clinton A. Bush, who had passed out from lack of oxygen and fallen through the open bomb bay, woke up on the ground with a fully extended chute, but he could not move. He had received a separated pelvis from the jolt when the already opened chute, cradled in his arms, completely deployed as he left the bomb bay and hit the slip-stream.[15]

On 24 October 1943, 10 days after this mission, the father of the left waist gunner, William Fink of Vancouver, Washington, received a second telegram from the War Department. The first one, delivered in mid-August of 1943, informed him his son, Staff Sergeant Loren M. Fink, had been wounded in action, over Kiel, Germany, on 29 July 1943. This second "wire" gave Mr. Fink the worst tragic news – his son was reported to have been "killed in action" on 14 October 1943. Sergeant Fink, who had fainted from the loss of blood just as he fell out of the waist exit of the plane, never had a chance to pull the rip cord of his parachute.[16]

In late November, 1943, William Fink opened a third communication from the War Department. This one finally contained some good news. His son's status was now changed from, "killed in action," to "wounded, and a prisoner of Germany"! Somehow, someway his parachute had opened during his descent. Loren Fink regained consciousness on the ground surrounded by German soldiers.[17]

Flight Officer Verl D. Fisher, the pilot, plus Technical Sergeants Clinton L. Bitton and Harvey Bennett, the engineer and radio operator, and tail gunner, Staff Sergeant George G. LeFebre, were all captured shortly after reaching the ground. All of them spent the balance of the war as prisoners of the Germans.[18]

After capture, Lieutenant Bush and Sergeant Fink were hospitalized for 12 days in a civilian hospital located at Geilenkirchen, Germany, just five miles from where their B-17 crashed. However, they received no medical treatment during that period of time.[19]

These two crewmen were then transferred to a French prison hospital at Dusseldorf, Germany. They arrived there on 30 October 1943, and both were quite ill. Fink had a severe infection and was treated by a Polish doctor. The physician told the waist gunner he had arrived at the hospital just in time or he would not have survived! The institution had no anesthetic for use during surgery, so both flyers received treatment without it![20] They left Dusseldorf on 21 January 1944, for the Frankfurt Interrogation Center, where they were separated and put into solitary confinement. They never saw each other again.[21]

After spending a week at Frankfurt, Sergeant Fink and many other Allied prisoners were loaded into boxcars for shipment to a prison located at Heydekrug, Lithuania. This camp, located several miles from where the Nemunas river empties into the Courland Lagoon (Kurisches Haff), was adjacent to the Baltic sea.[22] Fink, still suffering from his wounds, was now living some 60 miles inside the Soviet Union. He arrived at this, his first prison camp, Stalag Luft 6, on 2 February 1944, after six days of travel.

Due to the fluid front line fighting situation between the Russian and Germany armies, on 15 July 1944, he and the other prisoners were moved by boxcars 30 miles further to the north to the port of Memel (Klaipeda). Upon arrival at Memel, the prisoners were packed into the hold of a ship, and spent 49 hours sailing west through the Baltic sea. They disembarked at Swinemünde (Swinoujscie), a large seaport, just inside Germany where the Polish-German borders meet on the Baltic. From there, they then were moved by train farther east into Poland to Stalag Luft 4. This camp was located at a place called Gross Tychow (Tychery) and was in the middle of nowhere. The nearest inhabited area was the small town of Bialogard, some 12 miles to the northwest. The city of Kolobrzeg, situated on the Baltic Sea, lay another 20 miles to the northwest.[23]

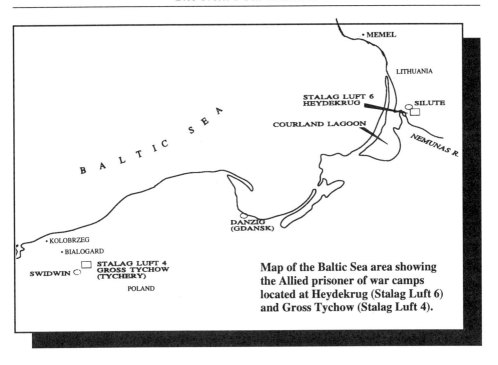

Map of the Baltic Sea area showing the Allied prisoner of war camps located at Heydekrug (Stalag Luft 6) and Gross Tychow (Stalag Luft 4).

The Exodus from Gross Tychow, Poland.[24]

Due to the collapsing German armies and their retrograde movements from the Russian front, Sergeant Loren Fink and thousands of his fellow prisoners moved out of Gross Tychow on 6 February 1945. This time the only transport available was their feet – so they walked. There was little to eat, and it was difficult to stay warm in the cold winds, ice, and snow of one of the worst European winters in memory.

They began the march moving due west and passed back into Swinemünde where many of them had docked some seven months earlier. However, this time there was no crowded ship waiting for the prisoners. They continued to walk in a westerly direction, passed through Swinemünde, and then turned southwest and headed for Neubrandenburg.

From there, the weary Allied prisoners continued southwest until they reached Uelzen, a small town some 58 miles northeast of the large city of Hannover.[25] Here at Uelzen the prisoners got a break from their constant walking. They were now loaded into railroad cars and began a train ride heading south and east. The train passed through the small town of Stendal, then turned due south and headed for the

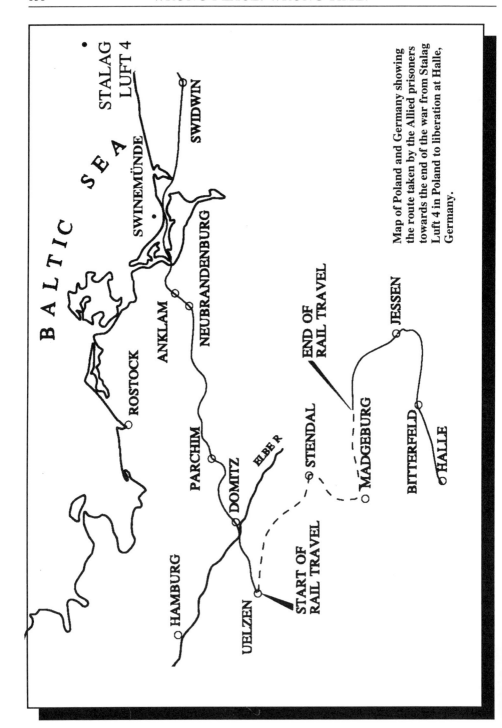

Map of Poland and Germany showing the route taken by the Allied prisoners towards the end of the war from Stalag Luft 4 in Poland to liberation at Halle, Germany.

large city of Magdeburg. After traveling by rail for some 90 miles, the train reached the vicinity of Magdeburg.[26] Just prior to entering Magdeburg, the train was shunted onto a short line, and it now continued on an easterly heading for another 30 miles.[27] The locomotive finally stopped when it ran out of track between the villages of Loburg and Wiesenburg.[28]

There was a brief respite for several days in the vicinity of the nearby village of Mockern. This area was called, "Tent City", and was also known as Camp 11A by the GIs.

After leaving Camp 11A, the prisoners continued marching east for another 20 miles, then turned southeast and passed through Jessen and Torgau. The marchers then left Torgau and headed due west to Bitterfeld. From there, they turned southwest and walked their final 18 miles to the large city of Halle.

It was here, on 25 April 1945, they met the vanguard of the American Army, and for these determined and resilient captured men the war was finally over! They had marched and shuffled as best they could for some 80 days, and covered approximately 559 miles. Ironically, this forced march was just 16 miles less than the briefed distance flown on 14 October 1943 by the bombers returning from their target. Schweinfurt to Beachy Head, England, the "coast in" point for the bombers was 575 miles.[29]

Chapter 15

The Following Eleven Minutes

There were now a total of four aircraft remaining from both the high and low squadrons, and all were stragglers! Kenyon, the only remaining ship from the low squadron, was still airborne while Farrell, Lang, and Skerry remained from the high squadron. All of them were under constant attack, and three of these ships would be destroyed during the next 11 minutes, 1351-1402 hours.

By pulling excessive power Lieutenant Farrell, the high squadron leader, had managed to stay fairly close to Normand's lead squadron, but was still racing to catch up and get in formation.[1] However, the other two ships behind Farrell could not keep up with him. Lieutenant Lang was on Farrell's left but well behind, as was the Skerry aircraft. German fighters continued to devastate the straggling American bombers of the 1st Division.

The Lang Crew

The Lang crew began bailing out of their ship at approximately 21,000 feet.[2] Staff Sergeant Steve Krawczynski, the tail gunner, was the last one out of the aircraft. He did not pop his chute until he was well below the others. At around 13,000 feet, he decided it was time. The chute performed as designed, and he was the first of the crew to hit the ground.[3]

The left turn made by the pilot, Lieutenant Robert S. Lang, just prior to his issuing the order to bail out was a blessing for several of the crew members. It put the aircraft much closer to Holland. As a result, at 1352 hours, Lieutenant Lang's empty B-17 crashed and exploded near the little German village of Puffendorf some 14 miles northeast of Aachen.[4] This crash site was just three miles southeast of where the Holt B-17 had crashed just two minutes earlier.

As they descended into the narrow Dutch corridor which lies between Belgium and Germany, Lieutenant Stanley Alukonis, the copilot, Staff Sergeant Howard J. Keenan, the right waist gunner, and tail gunner Krawczynski were about to fall into friendly hands.[5]

On his way down, Staff Sergeant Charles J. Groeninger, the left waist gunner, was buzzed twice by the same Bf 109. On his second pass, the pilot came so close the gunner could plainly make out the face of the pilot. The turbulence from the path of the aircraft caused the GIs parachute to partially collapse. As Groeninger scrambled to redeploy his chute and fill the canopy with air, he lost his gloves and oxygen mask. He landed safely on the ground next to a highway, and as he stood removing his harness a bus stopped nearby and emptied its Dutch passengers. They all ran to him and asked the American when would the invasion come? Before he could give any sort of an answer, he was captured.[6]

The ball turret gunner, Staff Sergeant Kenneth A. Maynard, landed on his back in an open field and could not get up and walk. The jolt of the opening of his parachute together with loose harness straps wrenched his lower body section. Technical Sergeant Warren E. McConnell, the radio operator, sprained an ankle on touchdown.[7]

The bombardier, Lieutenant James G. Adcox, hit back first, and for a minute thought he had broken it. Fortunately, he had not.[8]

After the navigator, Lieutenant John C. Tew, Jr., bailed out through the nose hatch he watched the tail of the aircraft go by, then he pulled the "D-ring." His chute, ripped and torn by flak damage, opened and filled with air. Tew looked up apprehensively at the fully extended canopy and could see a lot of daylight through the lacerated fabric. He noticed he was descending much more rapidly than the people around him, but the parachute was holding together. The sky was filled with parachutes. As he drifted earthward, Tew could see between 50-60 parachutes in the air at the same time. Most of the chutists appeared to be Americans from the 1st Division.[9]

Just before landing Jack Tew oscillated up against the side of a brick house, then fell into the back yard of the residence, injuring his head and back. He was still on the ground struggling to get out of the harness when the local authorities arrived to place him under arrest. As they marched him to a waiting truck, he had trouble walking. He looked down at his feet and realized while on the way down he and his right boot had parted company. Just before he had jumped, his wrist watch indicated 1348 hours. Again he checked his watch – it was 1400. He had been in the air about 12 minutes.[10]

Staff Sergeant Reuben B. Almquist, the engineer, together with Maynard, McConnell, Adcox, and Tew were all captured within minutes of each other on the outskirts of Maastricht, Holland.[11]

The local German authorities had watched, waited, and chased after as many of the descending American flyers as possible, but they would miss a few! With six of the Lang crew captured, the local German commanders kept their men looking for four more unaccounted flyers from this outfit. They were Lieutenants Lang and Alukonis and Sergeants Keenan and Krawczynski. The German authorities found Lang later in the day, but they would have a difficult time finding the other three.

Lieutenants Tew and Adcox spent approximately the next 19 months in Stalag Luft 3.[12] Sergeants McConnell, Maynard, Groeninger, and Almquist were incarcerated in Stalag 17B for the duration of the war.[13]

Lang's crew never saw him again. Shortly after their capture, a German officer reported to the crew one of their officers was dead. Later a German army doctor told one of the prisoners he had attended the funeral of an American officer whose description fit that of Robert Lang.[14]

The crew speculated any number of things could have happened to him. One of them figured he had been mortally wounded in the plane and never bailed out. Another crew member figured Lang may have been beaten to death by civilians after landing, while still another felt his pilot had been shot trying to escape.[15]

Lang's bombardier, Lieutenant Adcox, stated after he bailed out, ". . . I could see Lang's chute and we waved to each other. I turned to count chutes. We were all at about the same height. Unaware of the lack of oxygen, I couldn't count to ten. After several minutes, I finally saw nine chutes plus mine, so everything was OK . . ."[16]

In 1947, the mystery of what befell Lieutenant Robert S. Lang came to an end. On 15 February of that year, a letter was dispatched to the Lang family describing what had happened to their son. The letter was written by a German farm owner, Herr Josef Palmen, whose farm was located between Puffendorf and Loverich. It was endorsed by a German clergyman, Herr Hermanns, minister of Loverich and Alsdorf, District of Aachen.[17]

Farmer Palmen had been working with his people in a field located between the two villages of Puffendorf and Loverich on 14 October 1943. The air battle, high overhead, moved near his farm. A general alarm was sounded from the nearby village of Puffendorf, and he and his workers ran and hid beneath some hay stacks. They looked up and could see the many bombers and fighters struggling against each other. They also noticed numerous parachutes in the air from those who were falling out of contention.[18]

As they continued to gaze skyward, they witnessed a tragic and incredible occurrence. Approximately 165 yards from their positions, an airman slammed into the ground, and he was *not* wearing a parachute! Herr Palmen and his workers waited until the adversaries, still far above them, moved on and out of sight then they left their shelters. They immediately went to the spot where the flyer had landed. They found him lying on his back with no visible injuries but bleeding from his nose and mouth. The dead man was an American, and he was wearing an identification tag bearing the name, Robert Lang, San Francisco.[19]

Herr Palmen also stated in his letter to the Lang family another plane (the Holt aircraft) crashed nearby that day. He went on to explain the remains of Lieutenant Lang and those from the other crew were buried the following day in the Puffendorf cemetery.[20]

The Kenyon Crew

At approximately 1350 hours, the Kenyon crew had been under attack for some 17 minutes. Lieutenant Ellsworth H. Kenyon, the leader of the low squadron, now found himself all alone and in trouble.

He was about 21,000 feet with three engines on fire. He now gave the order for the crew to bail out.[21] Everyone but Technical Sergeant Russell R. Algren, the radio operator, left within moments.

Just before Staff Sergeant Arthur Englehardt, the ball turret gunner, bailed out through the waist exit he saw the radio room was on fire. He also saw Algren lying on the floor, and he appeared to be wounded. Englehardt tried to get to his friend, but could not due to the heat and flames.[22]

Lieutenants Thomas H. Davis, the copilot, and Joseph F. Collins, Jr., the bombardier, got away safely through the nose hatch.[23] They were followed by the navigator, Lieutenant John A. Cole, and then Kenyon, the pilot.[24]

Kenyon and Cole delayed pulling their rip cords, and did a free-fall. They did not have time to grab a portable oxygen bottle prior to jumping, and in order to get something to breathe, they had to get down to a lower altitude.[25]

The aircraft was now almost completely engulfed in fire. Back inside the aircraft, Russell Ahlgren somehow managed to find his way out of the burning radio section and put on his chute. As John Cole continued his free-fall, he looked up as Ahlgren bailed out of the burning B-17 opening his chute too soon. There was a bright flash as the material ignited from the flaming plane. The parachute flared briefly, then was consumed by the fire. As Cole continued to watch, the large flame went out and was followed by a wisp of smoke as the parachute vanished, and Sergeant Ahlgren fell some 20,000 feet to his death.[26]

Lieutenants Kenyon and Cole finally deployed their parachutes, and as they drifted downward in the clear weather both were buzzed by several fighters.[27] John Cole was too busy trying to survive an equipment problem to be concerned with the waving and saluting German fighter pilots flying so close to him. Earlier in the morning at Chelveston when the crew members were being issued their parachute harnesses, someone inadvertently picked his up and walked off with it. In a hurry, he drew another one but failed to adjust the harness straps so they would fit snugly. When he decided to end his free-fall and activate his chute, the jolt from the sudden opening of the canopy coupled with the loose harness caused the entire right side of the restraint apparatus to rip out. He now hung on for dear life with both arms wrapped about the left side of the harness.[28]

At approximately 1353, the low squadron ceased to exist! At this time, Kenyon's plane crashed and exploded between the German cities of Aachen and Duren. Kenyon's various crew members continued to float to earth between the two towns.

As Kenyon drifted down over the town of Aachen, he counted some 15-20 parachutes in the air. Shortly thereafter the pilot landed uninjured in a street of that city.[29]

Technical Sergeant Richard W. Lewis, the right waist gunner, came down near the edge of a small village and hit the ground hard as his chute partially hung up in a tree. He suffered a twisted knee and ankle and could not get to his feet. When the American landed, he was quickly surrounded by many Germans who thought he was a British flyer. They became very hostile towards him as they hated the British airmen for their random night bombing of targets that caused so many civilian casualties. After several minutes of heated discussion, Lewis was able to convince the unfriendly gathering he was, indeed, an American flyer. The Germans then calmed down and came forward to assist him.[30]

Lieutenant John Cole landed hard on his back still clinging to the left side of the parachute harness. He was dazed, his back hurt, and he was unable to get to his feet. While lying on the ground, he raised up on one elbow and saw he was surrounded by a group of farmers. He stared at them, and they stared right back. It was extremely quiet as neither the navigator nor the farmers uttered a word. They made no threatening gestures towards him nor did they try to harm him. However, every time he made a move, they moved with him. If he leaned to the right to try to get up, the group shuffled to the right. If he went left, they went left as they refused to break their circle around him. After several minutes of this cat-and-mouse play, Cole decided not to push his luck, and he simply sat perfectly still. A short time later the local police arrived, in a truck with a wood burning engine, and picked him up.[31]

The entire crew was quickly captured and reunited later that afternoon. The

German authorities confirmed the death of Ahlgren, the radio operator. His remains had been recovered, and the parachute he had been wearing was completely destroyed by fire.[32] Staff Sergeant Charles M. Green, the wounded right waist gunner, was detached from the crew and sent to a local hospital for treatment of his wounds.[33]

As with other downed Allied flyers, the remainder of the crew was transported to the Frankfurt Interrogation Center for prisoner of war processing. During the questioning of Lieutenant Cole, he continually refused to provide his interrogators with anything more than as required by the Geneva Convention; i.e., name, rank, and serial number. Finally, in an attempt to loosen Cole's tongue, the German major conducting the investigation showed the young flyer a map with the bomber route to and from the target marked with a red string. In addition, the major produced a book containing a complete roster of the 305th Bombardment Group which included the names of all the flight crews, commanding officer, adjutant, etc. The American continued to provide only his name, rank, and serial number. Finally the German interrogator departed, and the thought occurred to John Cole the enemy knew more about the 305th Bomb Group than he did.[34]

Later that evening as the four officers of the Kenyon crew sat in their adjacent prison cells, they were visited by a 21-22 year old German officer who claimed to be the Fw 190 fighter pilot who shot them down. He offered to take them to dinner, but the Americans declined.[35]

Lieutenants Kenyon, Davis, Cole, and Collins were later imprisoned in Stalag Luft 3, 7A, and 17. Technical Sergeants Finley J. Mercer, Jr. the engineer, Richard W. Lewis, left waist gunner, and Staff Sergeants Walter L. Gottshall, the tail gunner, and Arthur Englehardt, the ball turret gunner, were confined in Stalag 17B. After being treated for his wounds, Sergeant Green later joined his crew mates at 17B.[36]

The Skerry Crew

When he gave the order to bail out, Lieutenant Robert A. Skerry had run out of options. His B-17 had two engines shut down and feathered, one which was on fire, most of the tail surfaces were gone, and there was a large hole in the radio room. In addition, the bomber was being shot to pieces from both ends – flak was bursting all around the front of the ship while fighters were attacking from the rear.[37]

As he flew south over Bonn, Skerry waited a short time for the crew to clear the ship, then he left by the nose exit. The bomber was now down to 15,000 feet and continued to fly by itself, on automatic pilot, as it slowly lost altitude.[38]

The pilotless B-17 now headed in the general direction of the town of Adendorf as it continued to descend. Several minutes later, it crashed approximately 220 yards short of the village.

When it hit the ground, Skerry's aircraft was 29 miles behind Normand's lead squadron.[39] The remains of both Technical Sergeant Stanley H. Larrick, the radio operator, and Staff Sergeant Jack G. Johnson, the right waist gunner, were later recovered from the destroyed B-17.

The time, documented by six German antiaircraft units who all claimed credit for downing the B-17 and reported none of their fighters in the area, was 1402 hours. Later, Luftgaukommando VI, their higher headquarters awarded the downing of the ship to all six flak units. Not to be outgunned, Jagdgeschwader 26 "Schlageter" awarded a shot-down of a "Boeing" southwest of Bonn, at 1403 hours, to *Oberfeldwebel* Willi Roth.[40]

After Lieutenant John C. Lindquist, the copilot, bailed out he extended his free-fall and delayed opening his parachute for some time. On the way down, he decided to experiment before pulling the rip cord. He completely relaxed and fell like a leaf on its back. He then stiffened and arched his back and found he fell feet first. Next he assumed a diver's "jackknife" position, and he could look down as the earth came up to meet him. Finally, he did spins by using his hands as vanes and tilting them. He now noticed individual trees taking shape and decided it was time to use the chute. He pulled the rip cord, the canopy blossomed, and he was the first of the crew to touchdown. The copilot came down through tall trees, landing quite hard, and suffered a sprained back. As he began shedding his chute and harness he noticed his free-falling aerobatics had cost him his favorite wool scarf.[41]

In just a few moments he was greeted by many people from a nearby village. They peacefully motioned for him to come with them, and that evening he and the crew's left waist gunner, Staff Sergeant Gus Doumis, were reunited. They spent the night in a small village building – their first time ever in jail.[42]

After leaving the aircraft the navigator, Lieutenant Robert Guarini, waited a short time before activating his chute. He had been wearing gloves and using a "walk around" oxygen bottle when he jumped, but those items were now missing. The German fighters were also missing but not by much. As he drifted downward towards the ground, several of them flew very close to Guarini as they buzzed him. After several victory passes, they left.[43]

As he got lower, off to his left there appeared to be a small group of older German men blazing away at him with their rifles. Luckily for Bob Guarini the "home guards'" weapons were 1915 vintage, and their aim was poor. He hit the ground hard on one side and fractured his entire left rib cage. Some 15 minutes later he was in German hands.[44]

Skerry also delayed opening his chute to get to earth more quickly. He had an uneventful descent. However, he could not avoid landing in a tree. After clambering down this obstacle and finding a wooded area for cover, he was able to hide for about three hours until he was captured.[45]

When Staff Sergeant Wayne D. Rowlett was blown out the doorway of the bomber from the exploding cannon shell, he immediately pulled the "D-ring" and deployed his chute. He did not realize it, but his harness was loose and when the parachute opened he received a severe jolt and back sprain. The ball turret gunner was also greeted by numerous visits by three German fighters as he descended. He could clearly see the pilots in their cockpits as they flew by. He was not fired upon and eventually he, too, landed in a tree. After working his way out of his parachute and getting to the ground, he was able to evade the enemy for awhile. After five hours of freedom, the war ended for Dutch Rowlett as he was captured by the local authorities.[46]

After bailing out and immediately opening his chute Technical Sergeant Edwin E. DeVaul, the tail gunner, drifted far to the south. He landed in an apple orchard in a small village near Koblenz, some 16 miles from where he left his ship. He was captured by a young German paratrooper, Rainer Frieherr von Schlippenbach, who was home on leave and lived in the village. DeVaul was then taken to a hospital where he spent the next two months recovering from surgery for his wounds.[47]

The three officers of this crew, Lieutenants Skerry, Lindquist, and Guarini spent the remainder of the war together in Stalag Luft 3.[48] Staff Sergeant Robert J. Middleby, the engineer, together with Rowlett, Doumis, and DeVaul all were confined to Stalag 17B.[49]

Sergeant Cecil S. Key, the togglier, who refused many orders to bail out was never seen again by his crew members throughout the duration of the war.[50]

Throughout his stay in Stalag Luft 3, until his release in 1945, Lieutenant Guarini gave constant thought to the togglier. In a way, he felt responsible for the crewman's death. He continually reminded himself he should have forced the young sergeant out through the hatch, which would have saved his life.[51]

After the war, when the crew was reunited in France for the trip home, Bob Guarini saw his first ghost. It was that of Sergeant Cecil S. Key, or so he thought. After the rest of the crew had bailed out, Key finally decided to jump. He landed safely, but was captured shortly thereafter and imprisoned until the end of the war. Needless to say, Guarini and Key were both extremely happy to see one another.[52]

Chapter 16

Into Burg Adendorf [1]

After Lieutenant Skerry and his crew bailed out of their stricken ship, they never did see it again as it continued to fly by itself for sometime. Therefore, all of them were completely unaware of what happened to their B-17.[2]

In 1979, a 19 year old West German, by the name of Felix Freiherr von Loë, began a search to find the crew of a B-17 that crashed into the family home on 14 October 1943.[3] It took him some years to find out the serial number of the B-17 which was the key to opening the door to the story. By corresponding with many United States Government agencies, e.g., the National Archives, Washington, D.C., he was finally able to locate most of the surviving Skerry crew members.

It was not until Robert A. Skerry and six of his other crew members received letters from this young German gentleman in late fall of 1986, some 43 years after the plane crash, did they find out what had happened to their ship.[4]

The small village of Adendorf lies approximately eight miles southwest of Bonn, Germany, and about 40 miles due east of the Belgian border. Adjacent to Adendorf, just 220 yards from it, was a fortress dating back to the Middle Ages. This medieval stronghold was appropriately named Burg Adendorf.

Completely surrounded by a moat, it was only accessible by one bridge which in ancient times had been a drawbridge. The castle was "modernized" during the years 1659 and 1660, but some sections of the building dating back to the 14th century had been left untouched.

For many years the castle belonged to the von der Leyen family who came from the Rhine-Mosel region. In the 1820s, the aristocratic family of Count von Loë acquired the estate.

On 14 October 1943, this fort was home to seven persons. Among those was a 40 year old widow, Therese Freifrau von Loë, the owner, who ran the household. Her husband had died in 1938, and she was childless. The other six inhabitants of the castle were house-staff consisting of the housemaid, her elderly father, two of her sisters, and two other domestics.

The residents of this ancient estate were enjoying the sunshine and tranquility of this warm, autumn afternoon when the war came to pay them a visit. After lunch, the two sisters of the housemaid crossed the moat and departed for the vegetable fields adjacent to the estate. The sky was clear, and they could see for miles. They reached their work place and began harvesting beets.

They had been at work for about an hour when they heard a droning noise north of them. It was the three combat wings of the 1st Division with their tormentors, high in the sky, heading towards Schweinfurt. The women also heard many explosions and the rattle of machine gun fire from the running air battle. They stopped their work to look up and watch for a few moments. Eventually the bombers and fighters moved on and out of sight.

The women went back to harvesting but not for long! Almost as soon as they had started back to work, there were the sounds of more explosions, only these were much closer. They turned to look. It appeared to them a very high, stray bomber was being attacked by several fighters, and it was also receiving fire from antiaircraft batteries stationed around the city of Bonn.

The ship the two sisters saw was the abandoned Skerry aircraft. It continued to lose altitude as it past south of Bonn.

It was 1401 hours, and many persons in the town of Adendorf were busy following the progress of the descending, disabled bomber. Standing in the main street, one man noticed a light trail of smoke coming from the B-17. The smoke became darker, and then it turned to fire. As the plane continued to burn, those on the ground could see many parts falling from the ship.

The German fighter pilot, Willi Roth, who had been making numerous attacks on the B-17 from the six o'clock position now broke off contact with the crippled bomber. However, he continued to track it.

The plane continued to lose altitude, and now began a slow, shallow, right turn towards the Adendorf area. It appeared the ship was heading for Adendorf and suddenly there was fear among the people of the village the B-17 would crash into some part of their town. They began to scatter.

The two sisters in the field continued to stare at the stricken plane as it continued to descend. As it continued to settle, the disabled airplane now seemed to be heading straight for the two women. They quickly turned and glanced hastily about the open field for some kind of shelter – but found none. Wasting no time, they both began running back towards the castle. The bomber continued to draw closer to the fleeing pair and was but a short distance behind them. With the ship almost upon them, the sisters flung themselves to the ground. The burning aircraft past directly over them

Into Burg Adendorf

Above: Polish prisoners of war, supervised by German authorities, are shown uncovering the unexploded six-1000 pound bombs from the destroyed Skerry aircraft. Below: The wreckage of the Skerry aircraft lying in the courtyard of Burg Adendorf. (Both photos courtesy of Felix Freiherr von Loë)

Above: Present day aerial photo of Burg Adendorf showing the village of Adendorf at the top right. Below: Present day photo showing the entrance to the main building of Burg Adendorf. The demolished Skerry B-17 completely filled this courtyard. (Both photos courtesy of Felix Freiherr von Loë)

and now headed straight for the heart of the castle.

A civilian from Adendorf watched as the burning right wing of the B-17 buckled and broke off. At 1402 hours the aircraft fell to the right, disappeared over the tops of the buildings of Burg Adendorf and dropped into the courtyard of the castle.

There were three people occupying Burg Adendorf when the crash occurred. Therese Freifrau von Loë, the mistress, was awakened from her nap by the loud disturbance. Being only half awake, she thought the washing machine in the room upstairs had broken down again. However, when she looked out her window onto the scene in the courtyard, she realized what had happened. She quickly telephoned the local police and fire department. Both she and her housemaid decided to stay indoors.

The housemaid's father, who was living in the south wing of the servants' quarters, stepped outside to see what caused the deafening noise and could not believe his eyes. He was now staring at a large, four-engine bomber which was lying on the ground before him. From his position, he visually checked the confines of the castle. Somehow the plane had traversed several high structures of the estate, fallen into the outer courtyard, and had not touched any of the castle buildings.

The nose section of the B-17 was lying exactly in front of the stairs leading to the main house, with the broken and scattered fuselage just behind it. The tail section was resting on the opposite side of the courtyard within a few feet of a statue of the Holy Virgin. The remains of the plane completely filled the enclosed grounds of the estate. The bomber continued to burn and smoke after impact, and numerous rounds of .50 caliber machine gun ammunition began to explode.[5]

The buildings continued to withstand the heat of the fire and did not burn. However, two conifers and a red beech standing in front of the main entrance were not spared. They soon caught fire and were destroyed.

Fighting the fire proved to be a problem. Since the castle was surrounded by water, the fire brigade only had access to the castle by the one bridge. That, coupled with the fact many people thought the bomber might explode at any moment, caused the fire fighters to back off.

The local police arrived, took control, and no one was allowed to enter the castle grounds. Neither the police nor the occupants of the castle were aware there were six, hot 1000 pound bombs lying under the fuselage of the blazing ship.

The next day the fire burned itself out, and the remains of Staff Sergeant Jack G. Johnson, the right waist gunner, and those of Technical Sergeant Stanley H. Larrick, the radio operator, were recovered. Four days later they were buried in a temporary grave at the local cemetery in Adendorf.

About a week later the wreckage of the aircraft was removed from the grounds of the estate. At this time, the unexploded bombs were discovered. Fortunately, they had never been armed, since the crew was still early into the flight.

Therese Freifrau von Loë often prayed in the chapel of the castle to the Holy Virgin and Saint Florian, the patron saint against fire. After the fiery crash of the bomber in her courtyard, she was convinced she owed the salvation of the castle and those in it to both of them.[6]

Chapter 17

Then There Were Five

Major Normand continued to follow the 1st Combat Wing, flying the low group position, below and on the left of Lieutenant Colonel Milton's 91st Bomb Group. It was now 1422 hours, and Normand's four ship lead squadron was 10 minutes (some 40 miles) from Würzburg, the initial point (IP), and still intact. Up until this time the only two losses from his lead squadron had been the two bombers that were forced to abort before the battle began – the plane flown by Lieutenant R.E. Reid and the one piloted by Lieutenant Joseph E. Chely.

Also at this time Lieutenant Farrell's B-17, the only surviving one from the high squadron, pulled up high on the right alongside Normand's unit making a total of five 305th aircraft still in the fight.[1]

The 305th had not lost a ship for the past 20 minutes.[2] However, the Chelveston outfit still had about 18 more minutes to fly until it reached the target at Schweinfurt.[3]

The Maxwell Crew

Minutes from the IP Maxwell's aircraft, still in the #4 position behind and below Normand, came under heavy attack by a number of fighters. The ship caught fire and began to spin then recovered, and the pilot ordered the crew to bail out.[4]

Staff Sergeant William H. Connelley, the tail gunner, wasted no time and went out the tail escape hatch. After clearing the ship and deploying his parachute, he looked all around but did not see any other chutes in the air from his plane.[5]

Staff Sergeant Herbert S. Whitehead, at left waist, and Sergeant Charles H. Crane, the right waist gunner, moved quickly towards the waist exit. Whitehead went out first, and Crane prepared to follow.[6] At this time, the badly damaged aircraft blew up.[7]

It was 1429 hours when Maxwell's B-17 desintegrated. The plane's debris came down in the vicinity of Marionbrunn which was a small village 16 miles west northwest of the designated initial point of Würzburg.[8]

As Connelley drifted down, he looked for his aircraft, but it was gone. He did not see it blow up, but during his descent numerous parts of the plane floated past him, including one of the life rafts. He now saw another opened parachute above him, and he continued to watch it hoping it contained one of his crew members. He eventually landed near the other chute, and after discarding his harness and parachute he went over to the other flyer and found it was Charles Crane. The right waist gunner had been blown through the open waist door and was severely wounded and unconscious. Connelley began giving first aid to his injured crew mate and while doing so was captured by the German authorities.[9]

Sergeant Whitehead had deployed his chute as soon as he cleared the bomber. While descending towards the ground, he saw numerous pieces of his plane fall past him and concluded it had blown up. He was captured shortly after landing, and while in Dulag Luft in Frankfurt for interrogation, he was shown the dog tags of his pilot, Lieutenant Maxwell.[10]

Sergeants Crane, Connelley, and Whitehead spent the balance of the war as prisoners of the Germans in Stalag 17B.[11]

The remaining members of the Maxwell crew all died in their exploding aircraft. The four officers were Lieutenant Victor C. Maxwell, his copilot, Lieutenant Willis V. Rowan, the navigator, Lieutenant Urban H. Klister, and the bombardier, Lieutenant Andrew J. Zavar. The noncommissioned officers of the crew that perished were Technical Sergeants Silas W. Adamson, the engineer, and Jerome B. Pumo, the radio operator, together with the ball turret gunner, Staff Sergeant Craig T. Conley.[12]

The 305th, now down to four bombers, continued to follow Milton's 91st Group. They all reached the IP and made the turn at 1432 hours.[13] Schweinfurt lay dead ahead – eight minutes away. All bomb bay doors came open as the B-17s began their bomb run.

The Kincaid Crew

Lieutenant Alden C. Kincaid, still on Normand's right wing, had been wounded in the right arm by machine gun fire shortly after the air battle had begun. His copilot, Lieutenant Norman W. Smith, had been killed in the same attack and had been dead for almost an hour.[14]

At 1436, four minutes into the bomb run, Kincaid's aircraft took a number of direct hits from 20mm cannon fire. The #3 inboard engine caught fire and the outboard one, #4, began to lose power. The damaged aircraft would not respond to the pilot's controls as Kincaid could not maintain direction or altitude. He now ordered the crew to bail out. As everyone raced for an exit, the B-17 began to spin.[15]

The engineer, Staff Sergeant John F. Raines, snapped on his chest pack and started into the bomb bay to jump. However, he found the doors closed. He now managed to crawl into the nose section where he went out the nose hatch.[16] Lieutenant Robert D. Metcalf, the navigator, had already left as had Kincaid followed by his bombardier, Lieutenant Phillip A. Blasig. As Kincaid fell away from the plane, he saw the bombardier come through the hatch, and watched as Blasig was struck by the tail of the bomber.[17]

Both Staff Sergeant Kenneth R. Fenn, the ball turret gunner, and Sergeant Alfred C. Chalker, the right waist gunner were initially pinned down on the waist floor from the spinning plane. Fortunately they finally managed to exit the ship.[18]

Shortly after popping his chute, it felt to Sergeant William C. Heritage, the tail gunner, as if his feet were freezing. He looked down at them, and that was exactly what was happening. He had lost both boots and one sock.[19] The left waist gunner, Sergeant Louis Bridda, got away safely, but he, too, was suffering from cold feet from the loss of his footwear.[20]

As Al Chalker floated towards the ground, he together with Raines, Bridda, and Heritage were all buzzed by German fighters. However, the pilots did not fire at them. As they roared by, two of the Germans saluted and waggled their wings. Chalker and Heritage managed to wave back.[21]

Sergeant Heritage had several harrowing experiences on his descent. After he was buzzed, he found a pack of Camel cigarettes in one of his pockets with a book of matches inside the cellophane wrapper. He was about 10,000 feet up, low enough for the match to burn, so he decided to light up. Everything went fine. The match ignited, Bill Heritage drew on the cigarette, and the tobacco began to smoke and burn. He then tossed the lighted match up and away – and immediately had a sinking feeling. He looked up quickly and fortunately for him, the hot match had gone out and had not ignited the fabric canopy over his head. Later, as Heritage was about 200 feet above the terrain, he noticed a German male aiming a rifle at him. The German got in one shot and missed.[22]

As Fenn drifted downward, he saw Lieutenant Blasig falling with his chute still unopened, and he appeared to be pulling on the chest pack to get the parachute to open, but he never succeeded. As Kenneth Fenn continued to descend he could see many civilians and soldiers gathering on the ground. Fenn spoke and read German fluently, and he now noticed a sign on a small railroad building that read, "Kutzberg" – he was five miles west of Schweinfurt! He landed on his back in a rutabaga field, and was apprehended immediately. After capture, he realized he had been hit by flak several times in his right leg.[23]

At 1439, the Kincaid aircraft crashed approximately one mile southwest of Schweinfurt.[24] The bomber had almost made it to the target.

Lieutenant Smith, who had died early into the battle, went down with the ship as did the radio operator, Staff Sergeant Bernard T. Martin. Martin was apparently trapped inside the spinning aircraft and could not manage to make his way to an exit and jump. He was last heard on the intercom calling out fighters.[25]

Lieutenant Blasig's body was found lying within 500 feet of where the plane crashed. His chute was fastened to the harness, but it had never been deployed.[26] The next day, Metcalf and Heritage viewed Blasig's remains. No open wounds were visible, but the bombardier was badly bruised, and the entire right side of his face was crushed. Blasig, just as Kincaid, had been a former member of the Royal Air Force before the Americans entered the war.[27]

The other seven members of the crew survived the bail out. Lieutenants Kincaid and Metcalf were soon caught by the authorities and spent the remainder of the war in an officers' prison compound.

Raines was captured immediately after landing. Chalker hit the ground hard and was knocked unconscious. When he came to he was surrounded by Germans. Bridda was caught 15 minutes after touching down, while it took about three hours for the authorities to apprehend Heritage.[28] These four non-commissioned officers, together with the wounded Fenn, were imprisoned for the rest of the war in Stalag 17B at Krems, Austria, until 8 April 1945.

The Exodus from Stalag 17B[29]

On Saturday, 7 April 1945, a fever of excitement swept through Stalag 17B, the Allied prisoner of war camp at Krems, Austria. Staff Sergeant Kenneth R. Fenn and thousands of other inmates were alerted to prepare their packs for a long hike. They were told to expect one hot meal a day and lodgings somewhere each night. The Germans began by issuing the prisoners three Red Cross parcels for every five men. However, based on past experiences, the GIs took nothing for granted, and they prepared to carry all the food they could handle. In addition, some soap and shirts were taken along for trading purposes. Anything that could not be carried was either destroyed or given to other Allied prisoners, who were less fortunate.

The morning of 8 April dawned clear and warm. The Allied prisoners of war had been told the march would begin at 0700 hours, but the Germans had many problems getting the men to fall out. Many were sick, some were still healing from wounds, and they were all weak from undernourishment.

In Fenn's group, some 500 prisoners and 42 guards finally pulled out of the camp at 1000 hours. Most of the guards were Volkssturm or home guards. Being too old or disabled to fight as combatants, they had drawn the task of guarding ailing and worn-out prisoners of war (POWs).

It was tough going as the ragged band moved through winding mountain trails and past numerous fortified positions the Germans had prepared for the oncoming Russian units. The morale of the men was high, for they marched most of the day with the sun at their backs – they were heading west! After walking about 14 miles, they came to an apple orchard just outside the small village of Lobendorf. A halt was called and the men spent the evening there. There was no warm food or shelter available, and it was too cold to sleep. The men built fires and stayed close to the heat to keep warm. Twice during the night the guards, who now numbered about 100, had the fires extinguished due to Russian planes dropping flares in the area. No one wanted to be bombed, including the German guards, as all realized the war was winding down.

The next morning the group mustered to find everything covered with a light frost, and a heavy fog blanketed the area. Shortly thereafter the sun broke through the swirling mist, and it turned into another warm, sunny day. Today the hiking was easier on everyone as most of the terrain was downhill. The clusters of men moved generally west and covered another 14 miles before reaching the hamlet of Roxendorf. Here they assembled in a large open field and spent the night there trying to sleep on the ground. Again no warm food, but they were given bread – 18 men to a loaf. Fortunately, the captives had brought along their own food they had been constantly hoarding. The evening was again very cold, and few, if any, were able to doze off.

On Tuesday, 10 April, the officer in charge told his guards he had decided they would remain at their present bivouac for another day. There was an abundance of running water in the area, so many of the prisoners took advantage of another warm day to rest and take a much needed bath. The highlight of the this day occurred when the Germans dished out some cooked barley to everyone. Also included on the menu was a loaf of bread to be shared among nine men. The GIs, preparing for another cold evening, scrounged as much wood as possible to keep their fires alive throughout the night. Again, many slept on the cold ground.

April eleventh was another warm, sunny day and the tired crowd of men moved on passing through numerous small villages. Along the way they observed many run-down and abandoned farmhouses. The walking was easier now as the group began moving away from the rolling terrain and onto more level ground. Some of the prisoners, who had been in the sun for the past few days, found themselves picking up a slight suntan. The tired band of travelers succeeded in moving some 11 miles, and at dusk they found themselves near the small community of Laimbach. Things began to get better. The Germans served a small meal without bread, and some of the guards traded the GIs bread and potatoes for American soap and cigarettes. Many of the prisoners found shelter for the night in a number of nearby barns.

The next morning the prisoners awoke refreshed from their first good night's sleep since starting the march. Many had been able to sleep on straw found in the barns which was a respite from reposing on the cold ground. Today's journey began at noontime, and the band of guards and prisoners continued to move west in a steady rain. In order to keep warm, they walked as fast as they were able. By the end of the day, the tired group of men had moved another 14 miles west. They now arrived in the town of Ysperdorf. At mealtime, they received no food. However, that evening they were permitted to sleep indoors – this time in an empty paper mill. Sergeant Fenn, like so many others, traded for what he needed. In his case, he swapped some tea for eggs and bread.

On Friday the 13th, the rain finally stopped in the afternoon. The prisoners continued to plod along secondary roads and mountain paths as their guards made sure to keep them away from the busier routes and highways. Though the prisoners did not realize it, they were several miles north of the Danube and were paralleling the river as they moved generally westward towards the Austrian-German border. The group moved just over 14 miles which brought them to the village of Klamm. Before the war it was a town of about 2000 people, but when the prisoners arrived it was crowded with over 3000 refugees. Despite the many starving people in the area, the barter system worked – as long as people had things to trade. That day a bar of soap brought several eggs, some milk, and a couple of potatoes. The Germans passed out the only item on their bill of fare – hard biscuits. Many of the men found space in empty barns and were able to sleep indoors.

The rain began again the next day, and it was decided to keep the Allies at Klamm for another day. Numerous prisoners were now experiencing discomfort from hands, faces, and lips that were badly sunburned and chapped. At Klamm, many of the people were very friendly, and Kenneth Fenn's log indicated he ate well today. He recorded, "Rain again today but got 3 good meals. The people gave us a couple of old cows, but they were a treat . . ." The German entree for today was more hard biscuits. News reached the American prisoners their president, Franklin Delano

Roosevelt, had died two days earlier. Despite the wet and cold, some Schutzstaffel (SS) troopers arrived and ordered the GIs to put out their fires. The GIs protested! However, this brought a tirade from the Germans who wore the black uniforms, and the Americans reluctantly put out their fires. Fortunately, the SS that ruthlessly dominated authority in Germany and created terror throughout occupied Europe did no shooting today![30]

From Sunday, 15 April, through Tuesday, the 17th, the weather continued clear and cool as the arduous march kept on. The allied soldiers and airmen continued their journey to the west averaging about 14 miles per day during this period, as they moved over flat and picturesque terrain.

On Monday, 16 April, the worn-out group of prisoners, still heading west, approached the village of Mauthausen. A group of Jews, guarded by the SS, filed by them heading east. The SS had already shot and killed 18 of them and left them sprawled by the wayside. It was not difficult for Kenneth Fenn to see the remainder of the Jewish band was in a deplorable condition. The area around the town contained numerous bomb craters, and many unexploded bombs were lying about. The region was also littered with aircraft parts from downed planes. At mealtime, there was nothing to eat today!

The GIs also spent the next day, Tuesday, at Mauthausen. Everyone was now quite hungry, and it was nine men to a loaf of bread for today – the only food for the past two days. Some of the infamous SS troops paid the Americans a visit and apparently for the "fun of it" broke the nose of one of the GIs.

Food was becoming more and more scarce as the group moved out of Mauthausen and continued west on Wednesday, 18 April, and passed through the hamlet of Wilhering. The column of worn-out troops was now on its eleventh day of travel. After clearing Wilhering, a 25 mile walk, the marchers crossed the Danube, turned back east and walked another four miles. They now entered Linz. It was now pouring rain as the GIs entered this large city located on the south bank of the river. As they passed through this community an air raid alarm was sounding, so they saw very few people. That night everyone was quite miserable as they had to sleep in a soaking wet field on the outskirts of Linz.

On Thursday, 19 April the tired crowd of men kept on moving and covered a total of 15 miles during the day. The rain had stopped, but it remained cloudy and cool. In the afternoon, the haggard group of soldiers passed by a German airfield being manned by a group of Hungarians. Some 30 minutes after the prisoners passed by the airport, it was strafed and bombed by American P-38s. Fortunately, the prisoners had already moved on. At the end of the day, the tired band stopped at the village of

Hoisdorf. Here they were given cold rations to cook as best as they could, and five men shared a loaf of bread. Many of the Americans managed to find empty barns in which to sleep. Some of the buildings contained harvested wheat which the Americans cooked for several hours until it was somewhat palatable. They ate what they could of it just to stay alive!

The next day the sky cleared, but it continued to be cold. Today all of them traveled over dirt trails. It was slow going as most of the terrain rose in elevation. However, the prisoners managed to make over 16 miles – still heading west. They finally stopped near the village of Kallham. One meal was served which tasted so poorly many could not eat it. There was no bread. One fortunate thing occurred for the Americans – they again found barns in which to sleep.

The POWs also spent the next day, Saturday, in bivouac near Kallham. The Germans continued to issue something called "food." Today one meal was served, and again it was barely edible. For the second straight day there was no bread for anyone. In early evening, Sergeant Fenn, who spoke fluent German, managed to slip away to a nearby farmhouse and traded a khaki shirt for some potatoes, onions, eggs, and apples. Later that night a pleasant surprise occurred for the prisoners – they received one French, one Canadian, and one US parcel for each three men.

Sunday, 22 April, dawned cold and cloudy. It had rained all the previous night and now during the day a hard rain continued mixed with snow and hail. The Allies continued their march going uphill most of the day through heavy pine forests. It was extremely cold as they reached the higher elevations. Despite the weather and the excruciating ascent the fatigued men traveled over 14 miles. They finally reached a rather large town named Aurolzmunster. This day the Germans provided a decent meal, and a loaf of bread was issued for every five men. Fortunately, there were many barns in the area which provided shelter for the night.

The extreme cold weather continued with rain and snow on Monday. After moving west another 15 miles, the exhausted and hungry group arrived at Altheim, a town almost on the Austrian-German border. The GIs were served one meal today, and every 10 men shared a loaf of bread. There appeared to be numerous well-to-do farms in the area, but the locals refused to give or sell anything to the Americans. Again the POWs managed to sleep in the many outbuildings that dotted the area.

Everyone spent the next day here trying to rest and stay warm. There was little escape from the cold and rain which continued to plague the prisoners. One meal was furnished together with a loaf of bread for every seven men. Later in the day, some of the captives were able to slip out of camp. This time a number of the nearby farmers traded with the GIs.

When the Allies awoke on Wednesday, 25 April, they found the rain had stopped, but the ground was covered with frost. It was still extremely cold. The large gathering moved out and covered about 15 miles before reaching a large pine forest near Rossbach. The GIs were given nothing to eat today – not even bread. Many of them built huts out of pine boughs to make themselves as comfortable as possible. Fresh drinking water now had become an acute problem. The only way the prisoners could get any was to climb down very steep inclines inside the forest and trap the water as best they could as it trickled from small springs in the rocks. It was a difficult task for the weary and malnourished soldiers. Unknown to the Americans, they would remain at this same camp site for another week. The war was coming to an end, and the American liberators were not far away!

On Thursday, 26 April, heavy bombing, strafing, and artillery could be heard in the vicinity. There was now disorder developing among the German guards. The discipline of the guards had now become very bad unless there was an officer present. The guards and GIs now began sharing the same fires to keep warm. Today's meal consisted of some individual cold rations and 18 men to a loaf of bread. Late in the day, three French parcels were issued to every four men.

Friday morning brought more misery from the cold weather and hunger. A group of Russian prisoners, who were starving, began rioting for potatoes. Instead of potatoes, several of them were taken out and shot! The Americans fared somewhat better. They received more cold rations but no bread. The German guards were now also running out of food and had no shelter. A thunderstorm brought heavy rains later that night but the "pine straw" huts with moss covered roofs, the GIs had built several days earlier, kept the Americans dry.

It rained all day Saturday. Rations were very sparse with each man receiving several hard biscuits, a bit of salt and butter, and 25 grams of millet. One pine covered GI hut caught fire, and the one in which Fenn was staying collapsed from the heavy rains. Most of the prisoners were now soaking wet, cold, and very hungry.

Sunday morning, 29 April, arrived – cold and damp. Rumors were being spread around the camp the war was over! The prisoners tried to stay busy and warm. Many improved on their makeshift huts. Every man received three potatoes and one spoonful of oats. The morale of the German guards continued to deteriorate as their food was also running out.

On Monday, the last day of the month, rumors continued to circulate among the prisoners. The word now was the war would be over in a day or two! The weather remained cold and cloudy as everyone struggled to stay warm. Food for everyone was about gone. The ration today was a spoonful of beans and one of barley.

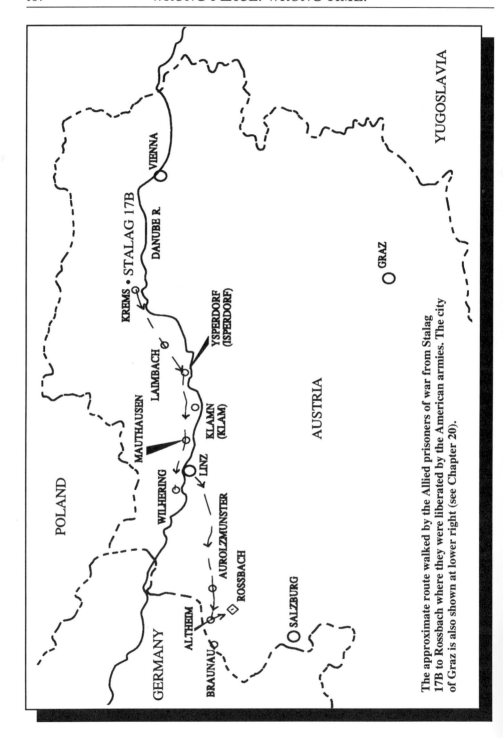

The approximate route walked by the Allied prisoners of war from Stalag 17B to Rossbach where they were liberated by the American armies. The city of Graz is also shown at lower right (see Chapter 20).

Throughout the day and that night heavy gunfire and explosions could be heard as the war drew closer to the temporary prisoner of war camp.

It was again cold and cloudy on the morning of 1 May. More gunfire was heard throughout the day. Today's ration for each man consisted of four potatoes, some beans, and one slice of bread. The latest rumor floating around the camp had the Americans surrounding the town of Braunau, the birth place of Hitler.[31]

Wednesday, 2 May 1945, began as another disconsolate day. The crowd of prisoners awoke to find the area completely covered with snow. Fenn reported in his log, "It looks nice, but it is damned cold!" The firewood was found to be wet, and it was very difficult to get a fire started. More rumors were heard in the camp: one, Hitler was reported dead, and two, Braunau had fallen to the Americans. One thing was certain to all the prisoners – there was very little food left for anyone! Everyone in camp hoped this was a good sign and the end must be near. Their hopes and prayers were answered at 1800 hours that evening. At that time, a captain with a reconnaissance unit from the 13th Armored Division, US Third Army, arrived on the scene. He explained to all the prisoners they would be freed shortly, but they had to stay in place for a few days.

The next morning, at 1130 hours, the American infantry arrived on the scene. The German guards surrendered and were marched off. Many of the POWs moved into the woods and rounded up several hundred more German deserters for the infantry to handle.

At 1800 hours, on Saturday, 5 May, the prisoners were issued "K rations", a lightweight package of emergency foods, and finally departed the Rossbach area. They now moved to a temporary American rest and rehabilitation (R&R) camp for several days, which included a mobile American Red Cross club which served stateside coffee and doughnuts.

On 9 May, Fenn's group was evacuated from a German airstrip by C-47 troop carrier aircraft for Le Havre, France. Here, with many other former American prisoners, he received further rehabilitation prior to returning home.

On 23 May 1945, Staff Sergeant Kenneth R. Fenn sailed from France for the United States and arrived in New York harbor on the morning of 3 June 1945.

Chapter 18

The Last Three

The Bullock Crew

With the Maxwell and Kincaid crews both gone, Major Normand's bomb group now consisted of himself with Farrell on his right wing and a burning B-17 off his left wing flown by Lieutenant Raymond P. Bullock.

The fighter attacks that destroyed the Kincaid aircraft had also caused problems for Bullock's plane. His ship was hit at 1434 hours by a 20mm cannon shell which tore into the fuel cell of the left wing. A small fire started, but soon most of the wing was on fire.[1]

Bullock still had six minutes to fly until reaching the target. Despite the knowledge his ship could blow up any moment, the pilot informed his crew they could bail out if they wished, but he was going to complete the bomb run then leave. The remainder of the crew all agreed to continue with the bomb run, and then after bombs away they, too, would bail out.[2]

The bombs were dropped at 1440.[3] As the three remaining 305th Bomb Group's B-17s came off the target, the Bullock crew members began to abandon their aircraft over the city of Schweinfurt.[4] The entire left wing of the plane was now engulfed in flames.

The last of the 305th bombers was about to fall to the enemy leaving two B-17s remaining from the group – Normand's piloted by Joe Kane, and the one flown by Barney Farrell.

All of the crew cleared the aircraft safely. They had departed the aircraft directly over Schweinfurt and with the clear weather and unlimited visibility all of them had a ringside view of the havoc occurring around and below them.

Copilot Lieutenant Homer L. Hocker, who exited the ship through the nose hatch, delayed pulling the rip cord. He finally opened his chute at about 5000 feet. He landed in a farmer's field and was uninjured. As he began shedding his parachute and harness, he noticed a middle-aged woman approaching him from a nearby farmhouse. It appeared to him she was carrying some sort of shotgun. He continued taking off the parachute harness as she approached to within 50 yards of him. He watched as she stopped, quickly raised the weapon and fired. The next thing Hocker knew he had been hit in the eye and left leg by some sort of "scatter gun." Despite his wounds, he took off running, but she did not follow. He continued across the field and soon lost sight of her. He crossed into another field and saw a number of unknown American flyers being prodded along by six German farmers, who were armed with rifles and shotguns. He limped over to the group and was impounded with the others. After being hospitalized for several days, he was transferred to Stalag Luft 3 for the remainder of the war. He never again saw any of his other crew members.[5]

Staff Sergeant Joseph K. Kocher, ball turret gunner, and Staff Sergeant Stanley J. Jarosynski, left waist gunner, had both jumped using the waist door. Joe Kocher let himself free fall for awhile, then pulled the rip cord. After his chute opened, his feet felt cold. Further inspection indicated he was shoeless. He had also lost a number of personal effects from an open zipper leg pocket – including his dog tags. As Kocher continued to drift downward, a German fighter passed by and the pilot waved his hand in greeting! Kocher landed in the tops of some trees in the orchard area of a farm. It took him about 20 minutes to free himself from his harness and climb down through the trees. When he arrived on the ground, a German farmer was waiting for him. The farmer began to manhandle him, and fortunately a German soldier arrived to take charge. He did not see any of his crew for several more weeks until he was confined to Stalag 17B, the prison camp at Krems, Austria.[6]

One day during his stay at the German stockade, 17B, Joe Kocher received a pleasant surprise. He was called into the office of one of the authorities and presented with his missing dog tags that had been lost during his parachute descent.[7]

Lieutenant Manning L. Lawrence, the navigator, and the togglier, Staff Sergeant Walter A. Kaeli were both captured shortly after landing. The tail gunner, Staff Sergeant Alden B. Curtis, and right waist gunner, Staff Sergeant Harold E. Coyne, were also caught by the authorities the same day.

Stanley Jarosynski also landed in a wooded area, but fortunately missed the trees. As soon as he was free of his parachute harness, he took off running. He had a compass and used it to point himself in a southerly direction towards Switzerland. He traveled as far as he could that first evening and then took refuge in an apple orchard. He had no food or water, so he subsisted on someone's apples and slept in an open field adjacent to a wooded area. As the sun began to rise the next morning, Stanley was again on the move. He was able to walk all that day without being seen. At dusk of the second day, he approached a farmer for help. Instead of assistance, the hostile German turned the American over to the authorities.[8]

Lieutenant Bullock, his radio operator, Staff Sergeant Harold E. Jackson, and Jarosynski were reunited in a local jail on the third day.[9] With the exception of the engineer, Technical Sergeant Carl Brunswick, who could not be found anywhere, these other members of the crew spent the remainder of the war in German prisoner of war camps.

Technical Sergeant Carl J. Brunswick[10]

When the Bullock crew began to abandon their burning B-17 Sergeant Brunswick, the engineer and top turret gunner, went out through the nose hatch and hesitated pulling the rip cord. He executed a "free-fall" to get as much distance as possible between himself and the plane. As he continued to fall, the airman noticed his helmet and left boot were now missing.

Brunswick came down between several trees just inside a forest and hit the ground very hard. As he struggled to his feet and got rid of the parachute, he realized his right hip was injured.

The engineer was also aware he was very close to the burning wreckage of his airplane, and he figured the Germans would soon be arriving on the scene. Limping badly he now took off looking for some concealment.

He immediately found a trail, and despite his sore hip he kept walking. A short time later he heard voices, and a squad of German soldiers came marching around a turn on the path several hundred yards ahead of him. They both saw each other at about the same time! Without delay, Brunswick hobbled off the footpath looking for a place to hide. He quickly found a large accumulation of leaves and fallen tree limbs. He immediately slid underneath the pile and covered himself as completely as possible. The flyer laid there as quietly as he could wondering when the searchers would arrive.

He did not have long to wait. In just a matter of moments, the soldiers appeared in the area. They moved about shouting back and forth to each other. Several times the Germans came within a few feet of his hiding place, but they did not discover him. Fortunately, he had picked a spot just off the trail, and most of the men from the squad had moved farther into the woods looking for the lone American. The soldiers searched until dusk, then gave up, and left the heavily wooded area.

After dark Sergeant Brunswick put his escape kit to use. He oriented his compass and started out on a course that would lead him to Switzerland. He also utilized the maps from the pack. He knew generally where he had come down and by correlating his charts with the road signs, he was able to maintain a course towards the Swiss border.

It was slow going with his injured hip, but he kept pushing ahead. He traveled from dusk until dawn and found places to hide during the daylight hours. Brunswick had very little to eat, and his foraging forced him to break into some isolated houses on several occasions. Most of the time he found nothing edible. However, during one of these break-ins, he came across an overcoat which he put to good use against the bitter cold weather.

After eight days of walking, on 22 October, the American flyer approached the city of Tuttlingen which was about 20 miles north of the Swiss border. His reconnaissance revealed the Danube river lay between himself and any further progress. He found the highway bridge over the river was heavily guarded, and his search for some sort of boat, for a one-man river crossing, was unsuccessful. He did find a railroad bridge not being guarded that led across the water, and he elected to try walking across this span.

Brunswick waited until well after dark. Around 0300 hours a thick, damp fog settled in the river bottom. He now moved out silently and slowly along the tracks that stretched across the river. After a short while, he began to make out the opposite riverbank through the mist. He kept walking and soon he could see the shore just ahead of him. He reached the opposite bank safely and stepped off the rails.

As he stood pondering his next move, a German sentry appeared out of the misty darkness in front of him and commanded, "Halt"! They were only 10 yards apart, and for a moment the weary and hurting airman thought about making a break for it. However, he could see the soldier was holding a machine gun pointed directly at him, so he decided to stay put.

Carl Brunswick's escape and evasion attempt was over. In about nine days, he had traveled over 200 miles wearing just one boot and with an injured hip. With little or nothing to eat, the American flyer had made it to within 20 miles of the Swiss border.

After his capture, he was placed in the prisoner of war pipeline via Stuttgart and the Frankfurt interrogation center. From there, he was shipped to Stalag 17B at Krems, Austria, where he spent the remainder of the war.

The Kane and Farrell Crews

These two crews were the only ones from the 305th Bomb Group that were able to reach the target and then manage to return to their base, at Chelveston, England. Kane landed at 1807 hours, and two minutes later Farrell followed him in at 1809.[11]

With the exception of Lieutenant Joseph W. Kane, Normand's pilot, no one else aboard that ship had been hit or wounded. Kane had experienced singed hair and minor facial bleeding from flying Plexiglas after a cockpit explosion.

Injury wise, Farrell's crew had not been as fortunate as Kane's. Just after leaving the target, Lieutenant Frederick E. Helmick, the bombardier, and Lieutenant Max Guber, the navigator, had both been severely wounded.

Both aircraft flown by these two crews had been badly damaged by enemy gunfire. The two B-17s were later restored to an airworthy condition and flew again, only to be lost in combat later in the war.

The 49th Station Hospital

Lieutenant Frederick E. Helmick

At the 49th Station Hospital, Lieutenant Helmick was recovering from his wounds to the left shoulder and back which required 56 stitches to close. While recuperating, the crew visited him, but he never had an opportunity to talk to others. After spending five days in the Diddington recovery ward, Helmick was returned to the 305th at Chelveston to pick up his personal belongings. He did not get to talk to anyone on the air base, and it seemed to him they were in a hurry to get him out of there! Later that same day, he was sent to a recovery hospital in Chorley, some 23 miles northeast of Liverpool. Lieutenant Guber, the navigator who owed his life to Helmick, was in intensive care for several weeks so the two of them never saw each other again. Helmick left England believing his navigator, Max Guber, had lost both eyes.[12]

On 27 April 1944, while stationed at Midland, Texas, Lieutenant Frederick E. Helmick was awarded the Distinguished Service Cross, the nation's second highest combat award, for his actions of 14 October 1943.[13]

Lieutenant Max Guber[14]

At the hospital, a few days following the raid, the doctor from the eye, ear, nose, and throat clinic (EENT) suggested to his clinic nurse, Lieutenant Bel Kaufman, she go visit one of his patients in the officers' ward. She was told he had a serious eye injury, and would most likely lose the sight in his right eye. In addition, he could lose the left eye due to a sympathetic hemorrhage. The doctor did not mention it was the Lieutenant from the 305th who he and the nurse had both met at the hospital on 7 October – a week before the Schweinfurt raid. Being very busy, the nurse forgot the doctor's request, and never did get to the ward to visit the patient. However, several weeks later she learned the doctor was talking about Max Guber.

Lieutenant Guber had come to the hospital on the seventh seeking medical help. He was having ear problems, manifested by pain and some loss of hearing associated with the altitude pressure changes of flying. When Max Guber entered the infirmary, both the doctor and nurse took immediate notice at the way Guber wore his cap. It made him look so different from the other flyers they had seen. His hat was pushed way back, as if he was about to take it off, and scratch his head. Besides, it was canted far too much to the left.

The navigator from the 305th was examined, given requests for sinus x-rays, a hearing test, and was scheduled for another appointment for 14 October. While completing the paperwork in the clinic office, Guber noticed the nurse, Lieutenant Kaufman, was wearing a pair of earrings. After several minutes of small talk, Guber mentioned to her he had never seen an Army woman in uniform wearing earrings. However, he told her he felt it certainly added a designer's touch to her nurse's outfit. She explained her occasional earring gig was simply a personal protest against Army conformity.

Lieutenant Guber did not appear at the clinic for his follow-up appointment on the 14th. In fact, there were so many no-shows it prompted concerned comments from many of the hospital staff members. They figured something big was going on that day, and they were right!

After some five or six weeks of recovery following extensive surgery, the wounded navigator was allowed to get out of bed and move about. One day, supporting himself on crutches, he hobbled into the EENT clinic. He had a bandaged eye and head, and was wearing the usual "issue" hospital robe, but this time, he was not wearing his hat. Clinging confidently to his props, he poked his nose into the clinic office and there sat nurse Kaufman. She recognized him, smiled and reminded him he had not kept his 14 October appointment. He laughed and told her he had arrived late but didn't remember how he got there!

From then on until he was transferred to Oxford General Hospital for consultation and continuing treatment, they shared a coffee break each day at three in the afternoon. In Britain, it was known as "tea time."

Oxford was quite a distance from Diddington, and Max and Bel did not see each other for some time. Several months later he returned to the 49th Hospital to visit her.

Lieutenant Max Guber and his flight cap that stirred the EENT clinic, 49th Station Hospital on 7 October 1943. (Photo courtesy of Bel Kaufman Guber, California)

He was now reassigned to another bomb group as "wing" navigator, temporarily placed on "duty not involving flying" (DNIF), and was located just 10 miles from the nurse with the earrings.

Since he was on DNIF status, and had many routine appointments at the 49th's EENT clinic, he managed a jeep ride almost every day to the hospital – and also to see Lieutenant Kaufman. Many afternoons she was able to take a break, and they met in the nearby town of Bedford.

They enjoyed each others company, but neither was concerned about getting involved, and making any lasting commitments. They both had their own dreams about the future, and these hopes seemed to go in different directions. Both young officers were independent, and planned to return to college and pursue their own interests when the war was finally over. Max wanted to return to the University of Wisconsin, where he already had a Masters degree, and use the GI Bill to further his education. Bel had her high school diploma plus her nurse's training but no degree. At the time, nurses were not eligible for any of the GI Bill benefits, so she was saving here money to go to college and get a Bachelor of Science degree in nursing. They both agreed their relationship was strictly an amicable one.

However, during these months of dating, they both began to realize their feelings were changing towards each other. Many things were becoming more meaningful when they were together – but still no commitment.

Lieutenant Max Guber, Farrell crew, together with Lieutenant Bel Kaufman, EENT clinic nurse, 49th Station Hospital. Photo was taken in 1944 prior to their engagement and wedding. (Photo courtesy of Bel Kaufman Guber, California)

In early January, 1945, they began making plans that included each other. They decided to marry, and the date was set for 12 June 1945.

When VE Day (Victory in Europe) arrived on 8 May 1945, signaling the end of the war in Europe, their plans had to change due to a sudden reassignment for Max. He was now ordered to attend a personnel school back in New York. Their wedding plans were quickly put on "hold." Max finally returned to England on 11 August, and this time all went well. The new wedding date was now set for 13 August 1945. They exchanged their wedding vows on that date. Later that day, they learned they were the first Americans ever married in Guild Hall which was one of London's oldest and most famous landmarks. The day after their wedding, 14 August 1945, there was even more good news. Hostilities had ended in the Pacific with the surrender of Japan, and VJ Day was proclaimed!

Both Max and Bel spent several more months in Europe after the war with the army of occupation. They returned to the United States in the fall of 1945, and fortunately, Max Guber's wounds healed. He suffered a modest right visual field loss in his right eye, but his left eye finally returned to normal.

Chapter 19

The Internees

The Dienhart Crew

At approximately 1349 the severely damaged Dienhart aircraft was at 21,300 feet when the pilot, Lieutenant Edward W. Dienhart, gave the order for his crew to abandon ship. At this time Dienhart did not know he had three badly wounded crew members on board.[1]

He knew the intercom was inoperative so he had activated the bail out alarm bell, and motioned excitedly to Lieutenant Brunson W. Bolin, the copilot, and Technical Sergeant George H. Blalock, Jr., the engineer, to bail out. They left their positions as ordered and exited the ship by the nose exit.[2]

Dienhart continued to fly for another minute to give the rest of the crew members an opportunity to get out of the plane. However, he was unaware the alarm bell was not working and the remaining crewmen had not received his message to leave the ship. Except for Bolin and Blalock, who were next to Dienhart in the cockpit and had just jumped as ordered, everyone else was still on aboard.[3]

After believing everyone had bailed out, the pilot decided not to jump but instead rolled the plane into a steep, diving turn to the left.[4] As the dive started the tail gunner, Staff Sergeant Bernard Segal, looked back up at the one other remaining aircraft still in the formation. He could see the Kenyon crew was also under heavy attack by a number of fighters. Segal released his tail escape hatch and put on his chute pack. He was not quite sure what was happening but was ready to bail out if necessary. He now hung his feet out through the hatch and waited.[5]

While sitting there, he put his headset back on and could hear Staff Sergeant Robert Cinibulk, at the right waist position, repeatedly calling over the intercom, "Waist gun to pilot, waist gun to pilot!" There was no reply from the pilot. Segal now figured everyone up front had either bailed out or had been killed. He decided to jump. As he turned he caught Robert Cinibulk's eye, and the waist gunner motioned for him not to jump – so he waited.[6]

The wounded and semiconscious radio operator, Technical Sergeant Hurley D. Smith, staggered into the waist section, and prepared to bail out. He was stopped by Cinibulk, who pointed to Smith's parachute. Hurley Smith's only escape device had a large hole through the center of it from a 20mm cannon shell. He had not realized it was unserviceable. Now he also waited.[7]

One enemy fighter pilot, who apparently observed the two parachutes leave Dienhart's ship, followed the aircraft for awhile as it continued to plunge earthward. However, the pilot broke off contact shortly thereafter apparently figuring the ship was going to crash.[8] At around 1500 feet above the ground, Segal was still sitting in the tail section with his feet hanging out the escape hatch and wondering whether the crew had a pilot or not. Just after that, the plane began to level out.[9] The pilot was now easing back on the yoke as he continued to let down. Dienhart was now several miles south of Cologne heading east as he crossed a large river. Though he did not realize it, he had just flown over the Rhine.[10] On the other side he leveled off just above the trees – finally rid of the fighters and flak.

Lieutenant Dienhart still believed everyone had bailed out, and he was all alone in the aircraft. He now turned to a southwest heading of about 230 degrees to get as far into southern France as possible. He figured he might be able to make it safely to Spain. Shortly thereafter he recrossed the Rhine and was now west of it.[11]

Several minutes went by when he felt someone tapping him on the leg from down in the hatch that led to the nose section. Dienhart was astounded when he looked down and saw Lieutenant Carl A. Johnson, the bombardier, was still on board. Johnson, who was badly wounded himself, immediately disappeared back into the nose section and soon emerged with another crew member. As gently as possible he eased the severely wounded navigator, Lieutenant Donald T. Rowley, up through the hatch and into the cockpit where he buckled Rowley into the empty copilot's seat. The navigator's left arm had almost been blown off and his right one was broken in two places. Despite the heavy bleeding and shock, Donald Rowley was conscious and coherent. He now found out Dienhart was heading for Spain. He told the pilot it was too far away and then gave him a heading of 180 degrees for Switzerland.[12]

Dienhart told Johnson to empty the bomb bay, and the wounded bombardier went back to the nose and salvoed the bombs in an open field. He tried to get the bomb bay doors to close, but they remained in the open position. He finally managed to shut the right one, but the other would not budge. He then returned to the cockpit, and the pilot asked him to go aft and see who else was on board. The bombardier told him he would try, but he was really hurting and he turned around to show Dienhart his wound – half his buttock had been shot away. Johnson gave up going aft, slid his way back down into the nose section, and laid down.[13]

After the pilot had pulled out of the dive at treetop level, those in the rear of the aircraft began to get active. Tail gunner Segal pulled his legs in from the escape hatch and went forward. He now joined Staff Sergeant Raymond C. Baus, the ball turret gunner, and the two waist gunners, Staff Sergeant Christy Zullo, and Cinibulk in the waist section of the ship.[14]

Segal and Cinibulk both made their way towards the cockpit. As they passed through the bomb bay, Bernie Segal saw the bombs were gone and one bomb bay door was still open.[15]

Christy Zullo, the left waist gunner, and Baus stayed with the badly wounded radio operator, Smith, and took care of his wounds. Zullo did his best to administer first aid, including a shot of morphine, to the wounded radio man who besides a broken arm, had some 75 pieces of shrapnel in his body. They then made Smith as comfortable as possible.[16]

Sergeants Cinibulk and Segal entered the cockpit area and found out the pilot was attempting to fly to Switzerland via a hedgehopping route over enemy terrain. The two gunners also saw Lieutenant Rowley strapped in the copilot's seat bleeding heavily from the wounds in both his arms. Dienhart sent Bernie Segal down into the nose compartment to render first aid to the bombardier and bring up all the maps he could find.[17]

In the meantime, Sergeant Cinibulk tended the wounds of the badly wounded navigator. Rowley had been given first aid and an injection of morphine earlier by the wounded bombardier to deaden the pain. However, he kept bleeding despite Cinibulk's best efforts to stop it.[18]

Down in the nose section, Segal found Johnson lying flat on his stomach and conscious. He treated Johnson's wounds despite working against the strong wind blowing through the shattered Plexiglas nose. He cut away the bombardier's parachute harness and Mae West and applied sulfa and dressings to several gaping wounds in Johnson's buttock. He then returned to the cockpit with some charts he found in the navigator's briefcase.[19]

After ministering to the wounded navigator, Robert Cinibulk crawled down into the nose to further help Carl Johnson. He found some more maps then proceeded to move the disabled bombardier from the nose of the aircraft, back through the bomb bay, and into the radio compartment. The trip across the narrow bomb bay catwalk was a hairy one for Sergeant Cinibulk. As he edged his way along the catwalk with one bomb bay door open and the ground under him just 25 feet away, he had to drag and half-carry Johnson sideways. Once in the radio room, Cinibulk wrapped Johnson in a parachute to keep him as warm and comfortable as possible. He then laid him down next to Hurley Smith.[20]

In the cockpit, Dienhart found the charts of no use as the ones of Germany had been blown out the shattered nose of the ship. Lieutenant Rowley was quite conscious, telling the pilot to correct his heading to 180 degrees in order to get to Switzerland. He was so alert, he kept complaining every time his pilot drifted off course. However, Dienhart could not hold a steady course as he threaded his way south through the western region of Germany. He had to zigzag as he followed the terrain staying low in the valleys trying to avoid radar and any possible flak towers. Dienhart had no idea of his location. However, as the ship flew its southerly heading through Germany at ground level, Rowley, despite his painful wounds, continued to insist on a 180 degree heading.[21]

In order to lighten the load and drag on the bomber, the four uninjured gunners began going through the ship tossing overboard everything that was not tied down. Among the expendables were such things as machine guns, ammunition, and radio equipment. To further reduce the drag on the airplane, the crew tried once again to close the one, opened bomb bay door, but they had no success.[22]

Lieutenant Dienhart continued to follow the valleys at treetop level. As he turned the corner of a small hill and leveled out, he looked up to see a German fighter, an Fw 190, about a quarter of a mile to the right front of him and several hundred feet higher than he. The fighter was now diving at the B-17 to make a head-on pass. Dienhart turned directly into the enemy plane. Now the fighter could not fire, since he could not get a good angle on the bomber without flying himself into the ground. The German pilot roared by, turned, and climbed to about 1000 feet. He now moved ahead of the B-17 in order to turn and make another frontal attack from a better angle. The Dienhart ship was now as low as an airplane could fly without hitting something on the earth's surface. The German fighter pilot began another pass from the right, diving as he came. The American pilot simply turned left into a convenient narrow valley putting a wooded ridge line between himself and the Focke-Wulf. The Fw 190 pilot pulled up and apparently in disgust flew off towards the north.[23]

Shortly after the German fighter departed the area, the bomber left the foothills and entered a wide, flat valley. Before Dienhart or the startled people on the ground knew what was happening, the B-17 roared across a large German airfield at an altitude of about 20-25 feet. It happened so suddenly the enemy could only respond by dashing for cover. Several parked aircraft were observed on the field, and not a shot was fired at the bomber. As quickly as the American plane arrived over the field, it just as rapidly left the enemy airport as it disappeared across the valley hugging the terrain. The B-17 once again entered the hills and continued on its southern journey.[24]

At approximately 1435 hours, the bomber crew was still on a southerly heading looking for a sanctuary, when it passed southeast of Saarbrücken, Germany. Moments later the Americans crossed the northeastern border that separates Germany from France. Dienhart was flying an excellent course, but he did not realize it at the time. He was now over France, just above the ground, paralleling the German border which was some 15-20 miles off his left to the east. The battered B-17 continued to stay in the valleys, hug the terrain, and fly south.[25]

The plane now passed a low range of mountains and entered a broad valley. At 1510, the crippled B-17 was still on the deck as it passed just alongside a French-Swiss border checkpoint at Rodersdorf, Switzerland.[26] Christy Zullo, standing in the waist, looked out the window and could see the border guards stationed at the gates and checkpoints that separated the two countries.[27]

Dienhart spotted a small town and began circling it, not sure exactly where he was – Germany or Switzerland! As he continued to circle trying to decide whether to land or not, the pilot eased the aircraft up to a more comfortable altitude of 500 feet. Rowley, still strapped in the copilot's seat, told his pilot they must be over Switzerland, and he was right. As Dienhart kept turning, people on the ground began waving at the Americans to come down. At the same time, someone in the town ran up a banner on the community flagstaff – it consisted of a white crusader's cross on a red field. Rowley, racked by his wounds, also saw the flag and said, "We're here, we are in Switzerland!."[28]

No doubt about it, they finally had arrived! The pilot now told the crew to prepare for a crash landing as he continued to circle, still trying to find a landing site. He could not go any higher as the weather in the valleys was deteriorating. Low clouds and fog were now obscuring much of the higher terrain as the ceiling and visibility began to drop. Dienhart kept looking for someplace to set the ship down, but every field he saw looked much too small.[29]

By now all the crew, except the pilot and navigator, had taken up their "crash landing positions" in the radio compartment. Lieutenant Dienhart kept flying and looking for a well-suited area, and he continued to come up empty. The mountainous terrain to his left and right offered nothing, and up ahead the hills were covered by mist and clouds.

At this time, 1533-1540 hours, the recent buzzing of a German airfield by a disabled four-engine American bomber had caused quite a stir in southwestern Germany. German radio traffic now indicated a number of their fighters were being employed to try and find a stray B-17 bomber which they thought was wandering around somewhere in Bavaria. Fortunately for Dienhart and his crew, the enemy was confused and looking in the wrong sector.[30]

Dienhart had been looking for almost 30 minutes for a place to land, and so far he had been unsuccessful. The weather continued to deteriorate, as the ceiling began to lower and the haze became worse. He estimated he had about 15 minutes fuel remaining. The pilot decided to double back for another look. He came around to the left, spotted a small town (Aesch) and headed for it. He was now six miles inside Switzerland.[31]

On the other side of the community, the pilot finally found what seemed to be an appropriate area for landing. It was a harvested potato field sitting on top of a hill.[32] It looked long enough and was into the wind. So with his landing gear up and one bomb bay door down, Ed Dienhart began to prepare to land. Due to the low ceiling and visibility, he flew a circular pattern so he would not lose sight of his landing area as he made his approach.[33] He started by flying upwind straight over the field, just passed the landing area, and then began a slow left turn. He continued his turn to the left through the upwind leg of the landing pattern playing the turn in order to get far enough out from his "runway." With that established, he kept on slowly turning to begin the downwind leg. He was now opposite the landing site. He made sure to extend this downwind leg longer than normal for he would need a lengthy final approach to land. Dienhart's circle now brought him around onto the base leg. From here, he could look off to his left and pick his landing spot.

He would be doing everything himself today to include: watching airspeed, extending flaps, and deciding the right time to shut down the three good engines before landing. The #1 and #2 engines were still functioning at full power, and the only way to slow them down was to shut them off. He eased back the throttle on #3 to slow the ship a bit.

The shallow left turn continued as the pilot came around on the last part of the circle. He now began to complete a 360 degree turn and to get lined up with his "runway" for the final approach for landing. The pilot had to be sure not to undershoot the field lest he and the crew end up at the bottom of the hill. If he misjudged and overshot the landing site, there would be no power available for a "go around" – it was now or never!

Lieutenant Dienhart had planned it well. He rolled out of the descending left turn, lined up with the field, and began his final approach. Rowley, still conscious, hurting and bleeding, watched from the right seat while those in the radio room hoped and prayed for the best.

As the aircraft descended and moved closer to the field, Dienhart approached

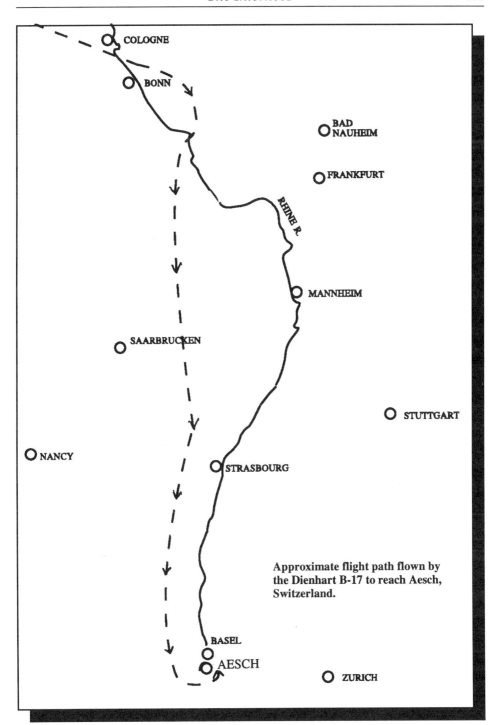

Approximate flight path flown by the Dienhart B-17 to reach Aesch, Switzerland.

what pilots call the "short final." He was now over the spot on the ground where he could make it to the landing area. He reached forward and quickly shut off the mixture control, and other switches to the three good engines, while dropping the nose slightly to pick up a bit more airspeed. As fuel starvation occurred in the engines, the propellers slowed and lost all power, but they continued to windmill for he did not have time to feather them. He saved the flaps for last, as extending them too early could cause him to undershoot the approach. As he neared the threshold, he reached for the flap switch to begin milking down the flaps – only nothing happened! Like so many other components of the airplane, they, too, were not working! With no flaps, the B-17 began to float, extending the touchdown spot well beyond where he had planned it. Rather than wait for the ship to slow itself and possibly overshoot the hill, Lieutenant Dienhart eased the bomber down and made contact with the ground.[34]

As the B-17 slid along the ground, the ball turret was sheared from under the belly of the aircraft "clean as a whistle", while the extended bomb bay door was ripped off. Dirt from the field began building up inside the bomb bay, acting as a brake.[35] The rest of the bomber was intact and finally slithered to a halt in the middle of the field which belonged to a farmer named Schlatthof.[36]

For several moments none of the crew members moved. In the cockpit, Dienhart and Rowley just sat. The plane had come to rest one mile from the Swiss town of Aesch which was six miles from the French border and five miles from the German one. The time was 1540 (1640 local Swiss time).[37]

Finally the navigator turned to the pilot and said, "Ed, that's the best landing you ever made"![38]

The rest of the uninjured crew began scrambling out the top hatch of the radio compartment, and as they did people began converging on the aircraft from all directions. Luckily there was no fire. The remaining crew members now began removing the badly wounded navigator, bombardier, and radio operator from the plane. Swiss soldiers and civilians arrived and offered their help. Soon two doctors arrived in an ambulance. Lieutenant Rowley, Lieutenant Johnson, and Sergeant Smith were checked by the physicians, given additional first aid, and then all three were transported to the nearest hospital at Basel for treatment.

The remainder of the crew picked up their personal belongings from the plane and were taken by the authorities to a nearby town just outside the city of Basel. There the crew was requested to present their identification which they did. The Americans were then fed bread, jam, coffee, and cake. The crew was also informed during the ongoing Swiss inspection of the their B-17, a 22 pound unexploded 88mm shell was

The Internees

Above: Members of the Dienhart crew standing on the wing of their B-17 after the crash landing inside Switzerland. Left to right: Raymond C. Baus, ball turret gunner; Bernard Segal, tail gunner; Robert Cinibulk, right waist gunner; Edward W. Dienhart, pilot; and Christy Zullo, left waist gunner. The three badly wounded crew members: Donald T. Rowley, navigator; Carl A. Johnson, bombardier; and Hurley D. Smith, radio operator, had already been removed to a nearby hospital by ambulance. To the right of Zullo is Dr. Huber, who treated the wounded and Mr. Maissen, who acted as interpreter. Below: The Dienhart B-17F, named "Lazy Baby", being guarded by a Swiss soldier after the crash near Aesch. (Both photos taken by Walter Höflinger, courtesy of Höflinger & Hauck, Switzerland)

found under the radio room floor in the camera well.[39]

Later that evening the Swiss authorities interrogated the crew, served them tea and rum, and then took them to a hotel where they met with General Bret Legg, the United States Military Attache.[40]

The glad news was the remaining eight crew members had made it safely to a neutral country, Switzerland, and were now considered "internees." The sad news was one of the eight, Lieutenant Donald F. Rowley, the crew's navigator, died from his wounds that same evening in the Swiss hospital located at Basel. The bad news was Lieutenant Brunson W. Bolin, the copilot, and Technical Sergeant George H. Blalock, Jr., the engineer, who were ordered to bail out by the pilot to save them from further danger, were captured and imprisoned by the Germans for the duration of the war.

Bolin suffered a cracked rib when his chute opened, and he managed to avoid a number of rifle shots fired at him while he was descending close to the ground. Luckily, the old German was a poor shot and was quickly disarmed by the two enemy soldiers who captured Lieutenant Bolin. Bolin and Sergeant Blalock, who also exited the aircraft safely, never did see each other again after the two Americans reached the ground.[41]

The Funeral of Lieutenant Donald T. Rowley[42]

Lieutenant Rowley's funeral was held at 1530 hours on the afternoon of 18 October 1943.[43] The five uninjured members of the Dienhart crew were gathered for this solemn occasion by the Swiss authorities and the American Military Attache at the Shuckanhauser Hotel in Basel. It was here the funeral motorcade began the trip to Hornli cemetery. The journey took the procession along the Rhine River, and adjacent to the German border. As the vehicles passed through the entrance to the cemetery, the German guards could be observed on their side of the border at the gate separating the two countries.

The group stopped in front of the church and was met by a large crowd of sympathetic Swiss civilians and numerous high ranking officers of the Swiss Army. They were also joined by many ranking officers from Poland, France, and England, who were internees like the Americans. The Swiss also provided a band and an honor guard. The Dienhart crew, all wearing black arm bands, entered the church and took their places alongside the casket of Lieutenant Rowley. Here they were joined by another American, from the attache's office, to make up the required six pall bearers.

After the church services were over, the six Americans picked up the casket, left the church, and fell in behind the honor guard and band. The large group then walked two blocks to the grave site where they were met by an English speaking minister who held the service.

After the services were over in the Hornli cemetery just outside Basel, Switzerland, Second Lieutenant Donald T. Rowley, United States Army Air Force, was accorded full military honors, and was laid to rest by his crew mates. It was an extremely impressive ceremony, and afterwards many wreaths were placed around the casket by both civilian and military personnel. Three volleys were fired over the casket by the rifles of the Swiss honor guard, while much to the crew's surprise, just across the border, the Germans fired three volleys from a cannon.

Lieutenant Rowley was later awarded the nation's third hightest combat award, the Silver Star, posthumously for "Gallantry in Action."[44]

The six pallbearers leaving the Swiss church with the casket containing the remains of Lieutenant Donald T. Rowley. Left to right: unknown U.S. military attache, Dienhart, Zullo, Segal, Baus, and Cinibulk. (Photo courtesy of Foto Jeck), Switzerland)

The Dienhart crew members, wearing black arm bands, beside the casket of their navigator, Lieutenant Rowley, just prior to burial at Hornli Cemetery, Basel, Switzerland. The Swiss army honor guard can be seen in the left background. (Photo courtesy of Foto Jeck), Switzerland)

The Palace Hotel, Adelboden, Switzerland[45]

The remaining Dienhart crew members, less Johnson and Smith, spent their first month in Bern, where they joined some 60 other internees from American bomber crews. From there Dienhart's crew members moved to their quarters at the Palace Hotel in Adelboden, a small village nestled in the Swiss Alps. Technical Sergeant Hurley D. Smith was confined to the hospital for five weeks of medical treatment while Lieutenant Carl L. Johnson spent the next three months recovering from his wounds. After their release from the hospital, they rejoined the rest of the crew at their civilian quarters at the Palace Hotel.

Allied military personnel seeking internment in Switzerland during World War II were well received and treated with dignity and friendliness by both the Swiss government and its citizens. The internees could come and go as they pleased. However, except for special permission to travel elsewhere, they were restricted to the valley containing the hotel that housed them.

From time to time they were given some of their pay in order to buy clothes and personal items. Most of them wore civilian clothes, usually consisting of sport and ski clothing, and if needed the International Red Cross provided them with a new issue of military clothing.

There was no curfew in effect – one was not needed. There was no night life in the nearby town, and the residents settled in early. It was a place where no one stayed out late at night. The Americans played a lot of cards and stayed close to the hotel.

Many of the GIs participated in the winter sports available in the area. For a lot of them it was the first time they had ever seen snow, and soon quite a few of them became adept at ice skating and skiing. There was an ice hockey rink in front of the hotel and with the help of the Red Cross providing uniforms and skates, the Americans formed a hockey team. They played many of the Swiss teams but had trouble finding a way to win against the more mature and experienced alpiners.

Several months after arriving in Adelboden, the Palace Hotel filled up with Americans, and Lieutenant Dienhart was placed in charge of another, smaller hotel housing Americans in the center of town. Shortly after Dienhart moved into his new quarters, all the inns were inspected by the American military attache, General Legg. It was at this time nine of the GIs, including Dienhart and his bombardier, Carl Johnson, were selected to attend school in Geneva. The remainder of the various crews remained in Adelboden.

While in Geneva, the American servicemen continued to live as civilians. They attended school in the morning, completed homework assignments in the afternoon, and spent their free time swimming and playing tennis. They also met a lot of congenial Swiss civilians and were invited to several homes for dinner. The flyers also became acquainted with the Swiss commandant of all internees, Major Mathy, and later they were introduced to his son.

The Paroles[46]

The Swiss government made life as pleasant as possible for the Allied internees. Nevertheless, it had certain laws of neutrality the "guests" were required to observe and obey.

Technically the Dienhart crew entered the neutral country of Switzerland bearing arms; i.e., the B-17 carried machine guns and ammunition while the crew members bore side arms. The Swiss law simply said if a military individual, seeking sanctuary from the Germans, entered the country bearing arms, he was considered an "internee" and must wait in Switzerland until the war was over before returning home. The internee was required to sign a parole he would not attempt to escape and leave the country. If one was caught breaking this parole, he was in for harsh treatment.

This law did not apply to many of the British interned in Switzerland. Much earlier in the war many of the British military captured fighting the Germans in North Africa were sent to prisoner of war camps in Italy. In September, 1943, when the Italian war machine and Mussolini's government collapsed, the Germans rushed to Italy to restore order, occupy the country, and continue the fighting. However, before the Germans could get in place, the Italian guards opened the doors to the camps, and the imprisoned British took off. Some of them opted to go south, while others decided to work their way north to Switzerland.

The British entered Switzerland not bearing arms, and were considered "escapees" and thus were free to leave whenever they could. However, no one wanted to leave Switzerland and run into the Germans, so everyone stayed put for awhile.

After the invasion of Europe on 6 June 1944, the situation for the Switzerland internees and escapees began to change. As the Allied armies slowly pushed their way east, they came closer and closer to the Swiss border. Finally they reached the edge of Switzerland and much of the border was occupied by the Free French.

Even though the war was not yet over, the British could now make plans to return home, but the Americans were forbidden to leave. Despite the unpleasantness and hardships associated with a failed escape attempt, Dienhart, Baus, and Segal went ahead and made plans to make an early attempt to leave – and one of them got caught.

October, 1944 was "getaway" month for many of the Allied internees. The Free French had occupied the French-Swiss border, and the British "escapees" were leaving in droves by train of their own free will.

On the other hand, if the Americans wanted to leave, they had to scheme and connive to find a way out without getting caught. Some of the GIs elected to stay put until the war's end, but many were tired of being spectators, wanted their freedom, and the opportunity to be able to return home.

In early October Major Mathy had his son check the border for a good crossing point for several of his American friends. When the timing was right, the Major's son led Lieutenant Dienhart and another American officer to the border, where they were met by the Free French. From there the two Americans were taken to an air base in France, flown to England, and then to the United States.[47]

On 4 October, Sergeants Segal and Baus elected to "go for it!" They each joined a different group of aspiring Americans and were not together when they made their escape attempts.

The Americans, who lived near and were friendly with the British in Adelboden, enlisted them to help with their plans. The good-natured British chaps agreed to assist their American "cousins" in any way they could.

In early afternoon, the Americans began arriving at the Grand Hotel from where the British would depart. The GIs were given complete Limey battle dress, bought green identification cards, obtained medical cards, and used false names – Baus chose, "Charles Case."

At 2100 hours, the subterfuge got underway. As the British boarded buses for the train station, Swiss guards checked identification cards. As the convoy proceeded along the road, Swiss guards could be seen positioned all along the route. The groups finally arrived at Frutigen, had tea, then boarded the trains.

Up ahead, Segal's bunch was already on its train when Baus' crowd began loading. The meticulous Swiss authorities were now ready with rosters of who should be aboard the trains.

In Segal's railcar, the authorities were impatient. Their head count did not tally correctly, and somehow their lists were inaccurate. After a recheck, those in charge left the train and had a brief meeting on the station platform. There was much arm waving, loud discussions, and finally a decision was reached. They would take one more roll call. They did, and there were still too many people on board in British uniforms with valid ID cards. The Swiss officer in charge threw up his hands in disgust, called his people off the train, and released the car for travel.

In Baus' coach, things were a bit different. One American was discovered, and he was removed from the train. Baus' green card passed inspection, and he breathed easier.

The train now departed for Geneva. The troops were fed and did some sleeping along the way. During a brief stop, IDs were checked again, and Baus' luck continued.

The train arrived in Geneva, and the group was taken by tram to the custom house. Bernie Segal made it through the tight inspection with his outfit and in three days was back in England. One week later he returned to the United States.

Subsequent checks of the other groups going through inspection became more thorough. Instead of moving through as a unit, the authorities began inspecting one or two of the men at a time. Another roll call was taken, and Ray Baus got by using someone else's name. However, he could not get past several authorities who were sitting at a table scrutinizing everybody who came their way. Baus backed off, left the building, and went outside to the men's room. It was here he and seven other Americans were arrested for attempting to escape.

Things now went from bad to worse. The eight captives were interrogated several times and spent the rest of the night in one jail cell. They were fed bread and soup and received no cigarettes – not a good omen.

The next morning they were sent to a formal prison camp, issued a blue fatigue outfit, mess kit, and assigned to a barracks. They now resided in a prison surrounded by three fences of barbed wire and ate the rations of condemned men.

This situation was a complete reversal from the conditions of living as an internee in a hotel. In the following weeks many other Americans were incarcerated in the prison for their unsuccessful escape attempts. A number of GIs tried to escape from the camp, but most were caught, returned, and placed in solitary confinement.

On 12 November 1944, a group of senior American officers, headed by General Legg, was permitted by the Swiss government to visit with the numerous American officer and noncommissioned officer prisoners. The inmates were offered a 30 day parole, and all agreed to sign it.

Two days later, after spending 42 days in a Swiss prison, Baus returned to his hotel in Adelboden, and his rights as an internee were restored. Thereafter, from time to time, the Americans were required to sign temporary paroles. In March, 1945, Staff Sergeant Raymond C. Baus left his hotel in Switzerland and returned to the United States.

Sergeant Christy Zullo, who joined the Dienhart crew as a replacement for a sick crew member on 14 October 1943, spent 19 months as a guest of the Swiss

Government in Adelboden. He had a special reason for remaining an "internee" and not wanting to try to escape.[48]

Early in his stay, he and eight other airmen volunteered to help build an American cemetery in Münsingen. They commuted each day to the site, which was located about 35 miles from their hotel. To keep busy in the evenings, Zullo began participating in a Swiss custom observed by the natives of Adelboden. It was called an "evening stroll about town." It wasn't long before many of the other Americans soon picked up on this practice and joined in the nightly promenades.

One early evening in November, 1943, while out walking, Christy Zullo experienced something that would change the entire course of his life – he met a girl! Hilda Schranz, a beautiful, young Swiss lady was sauntering with her sister when the three of them met and introduced themselves. All were in their early twenties. Hilda's sister spoke fluent English, and it was not long before the three of them became acquainted. The American was then invited to walk with the two sisters. The more they walked and talked, the more Zullo fell in love with Hilda. Shortly after that their courtship began.

They did not have far to travel to see one another. They lived just a half a block apart – he with his comrades in the Nevada Palace hotel and she with her family.

Her father, Peter Schranz, owned and operated a store selling dairy products such as: milk, eggs, and cheese. Hilda worked in the store and helped her father with the business. The family quarters and the store were under the same roof much like the "mom and pop grocery stores" in the United States.

They decided to marry early the next year, but first by regulation the flyer had to have the approval of his commanding officer (CO). Since his CO was stationed at Chelveston, England with the 305th Bomb Group, obtaining permission to marry was a problem – but not for long.

Brigadier General Bret R. Legg quickly came to the rescue. The Military Attache at the American Embassy in Bern used his authority to approve the wedding plans.

On Sunday, 7 January 1944, Staff Sergeant Christy Zullo, United States Army Air Force, and Hilda Schranz of the Federal Republic of Switzerland became man and wife. They were married in the Protestant Church of Adelboden attended by a small family gathering. This wedding, between an American flyer and a Swiss maiden during wartime, added even more luster to the history of this sacred place of worship, which dated back to the year 1433.[49]

Zullo returned home in April, 1945, and his wife, Hilda, received permission to come to the United States and join him one year later in April, 1946.

The other three members of the Dienhart crew living in Switzerland, Lieutenant

Carl A. Johnson, Technical Sergeant Hurley D. Smith, and Staff Sergeant Robert Cinibulk, also returned home safely to the United States after the war ended.

Lieutenant Brunson W. Bolin and Technical Sergeant George H. Blalock, Jr., who had been hastily ordered to bail out over Germany and had become prisoners of war, were released from capture at the end of hostilities and returned to the United States.

Chapter 20

The Escapees

Staff Sergeant Herman E. Molen[1]

After capture Herman Molen, Eakle's togglier, was placed with many other 305th personnel in the German prisoner of war camp, known as Stalag 17B, located at Krems, Austria. In just a short period of time he, like so many other Americans, became restless and wanted out of the place.

On Easter Sunday, 1944, he and another American attempted to escape from the camp. They decided to go it alone and not tell anyone. Therefore, it was an uncoordinated escape in that the two Americans did not clear it with the "Escape Committee."

This committee consisted of a group of prisoners headed by the senior inmate of the prison to plan, establish procedures, and coordinate escape efforts for comrades who wanted to "break out." Most successful escapes occurred when this self-constituted organization was included in the "getaway" plans for, among other things, it could provide maps, uniforms, money, food, diversions, guidance, and prayers.

Molen and his friend had a certain area of barbed wire fencing under surveillance for some time. The barricade had a watch tower at each end, and it was patrolled by one sentry.

On the morning of the feast that commemorates the resurrection of Christ, the flyers cut their way through the wire in broad daylight and took off. They were captured one week later and were returned to their barracks at Krems. They were not physically abused but did receive a punishment of 21 days on bread and water.

In the early summer of 1944, Herman Molen and a number of other American prisoners began planning another escape attempt. This time the togglier knew better and he coordinated his efforts and worked together with the "committee."

Stalag 17B was divided into two camps. One contained the American prisoners, while the other held a mix of Allied personnel such as: Australians, New Zealanders, and English. Thus, the GIs dubbed it, "Camp International."

The Americans and the Allies across the way from them kept open excellent lines of communications. Using various ways of moving messages back and forth, both groups had the latest word on what was going on!

Among the rumors floating around both prisons was one concerning "Camp International." Based on a reliable source, the "word" was a group from the International side was going to be transferred to Graz, Austria to a "work and labor camp."

Being in a "work camp" rather than a "prison camp" appealed to Herman Molen for several reasons. One, there was no imprisonment; and two, it provided a chance to just walk away from a work detail during the day – and escape!

The committee provided Molen and eight other GIs with chocolate and cigarettes with which they used to bribe selected German guards. In turn, the guards allowed the Americans to visit Camp International.

During one get-together, the nine Americans traded identities, including uniforms, with an equal number of Allied prisoners. When the visit was over and it was time to go home, the GIs remained at "International" while the mix of Aussies, New Zealanders, and Englishmen were escorted to their new quarters in the American camp.

After a brief stay with the Allies, Molen and another American were able to wangle themselves in among those destined to be shipped to the work camp at Graz. They soon arrived in their new surroundings where they were assigned outside toil such as: road grading, chopping wood, and any other details the Germans required.

One morning while Herman Molen and his buddy were clearing brush alongside a dirt road, a young, attractive, and buxom girl came down the route leading an overaged horse saddled with firewood. The lass caught the wandering eye of the lone German guard and she gave him a big smile. He turned to the two Americans, told them to stay put, and for them to continue their work. He then proceeded out onto the road to talk to the pretty damsel. He offered her some candy, which she accepted, and then they both began a lighthearted conversation.

As the two flyers continued to work, they occasionally glanced back at the German soldier as he continued chatting with the maiden. Now and then the airmen would move ever so slowly away from both the sentry and the young lady. The guard would occasionally look over at the two GIs to check on them, then he finally became completely engrossed with the girl. As they continued to perform their work, the two prisoners edged their way farther down the road. Sergeant "Moe" Molen, who had been watching his guardian out of the corner of his eye, was now convinced it was time to make his move. He motioned for his buddy to follow him, and the two of them walked slowly into the woods and disappeared. Much later, to his misfortune, the soldier realized his two charges were AWOL.

The two American escapees had already made their plans. They knew Graz was only about 30 miles northwest of the Yugoslavian border, so they now headed southeast towards the highlands of that country.

After several days of travel, they crossed the border into Yugoslavia and began their climb into the mountains. It was not long until they were captured by one of the many small bands of Tito's Yugoslavian Partisans. This guerilla party was led by a Russian officer, who mistrusted Americans. He informed the two flyers if they tried to escape they would be shot.

The suspicious Russian never allowed the GIs to carry any weapons. Their talents were used primarily as porters and laborers in the rugged, mountainous terrain, carrying such things as ammunition and other supplies.

During their stay with the partisans, the Russian leader made one exception to his rule of, "Thou shall not bear arms!" Early one afternoon, the ragtag group was ambushed by a German mountain unit, and a firefight broke out. After about 20 minutes of exchanging gunfire, it was obvious to the Russian he and his men were pinned down, and they were also taking a pounding. He needed more firepower, and right now!

To the surprise of the two well concealed and safely entrenched Americans, they were supplied with automatic weapons, bandoliers of ammunition, and hand grenades. The Russian ordered the Americans to move out to their right and engage the enemy. Both airmen quickly joined the fight, and the Germans thinking they were being outflanked withdrew to consolidate their positions. While doing this, the guerillas and the two B-17 Gunners slipped away.

Much later when the partisan group reach an area of safety, the Russian called a halt and approached the two Americans. Both gunners figured they now had earned the respect and trust of their leader and would be accepted as members of the team. Instead, the officer pointed his pistol at them and ordered the Americans to drop their weapons. Once again Herman Molen and his friend were back in the "porter" business.

The two GIs spent about a month with this outfit, and one day a British Army Intelligence officer parachuted into the camp for consultation and coordination of future guerila activities. The Americans were allowed to converse with the British officer, and he placed the disposition of the two airmen on the agenda to be discussed with the Russian. After much haggling, the Russian finally agreed the Americans would be released to make their own way as best they could.

One week later, the two gunners made their way out of the mountains and began walking the back roads through an area dotted with small farms. While resting along their route, they noticed a cart approaching them from a distance. It was drawn by a single, very tired, and undernourished horse. The wagon was stacked with a large amount of hay and was driven by a kindly looking old man.

As he approached, he waved to them and stopped the wagon. Molen and his friend got up from where they were sitting alongside the road and walked over to him. Both explained, as best they could, they were American flyers who had escaped from the Germans, and they needed help. All the time the airmen were talking the elderly man kept nodding his head up and down and smiling as if he completely understood their predicament. He seemed most sympathetic and finally offered to hide them from the enemy until they passed through the next village, which he explained was occupied by German troops. He also told them, once past the town, they would be in open and safe country again.

The Americans were delighted, climbed up on the rig, buried themselves under the pile of fodder, and relaxed as the buggy got under way. Some 30 minutes later the wagon stopped, and they could hear excited voices. Molen sneaked a peek out from under the hay. He observed the "friendly" farmer talking to several German soldiers at a check point and pointing excitedly to the pile of hay. The kind-hearted farmer had just turned them in, and Molen's second escape attempt had failed.

Sometime later he and his buddy were graciously welcomed back to Stalag 17B by the authorities who started them off in solitary confinement – this time with 30 days of bread and water! Herman Molen did not attempt a third escape try but decided to wait out the remainder of the war in the prison camp.

Staff Sergeant John L. Gudiatis[2]

The badly wounded right waist gunner of the Willis crew, John Gudiatis, had passed out near the waist door of the ship before he was able to jump. However, he came to just before hitting the ground and his parachute was fully deployed. He realized the other waist gunner, Staff Sergeant Edward J. Sedinger, was the only one who could have activated his chute, pushed him out the exit, and thereby saved his life. Sedinger, who refused to jump, went down with the ship.

After receiving emergency medical treatment for his numerous wounds Sergeant Gudiatis, like so many other injured American flyers, began the first of his many moves about German occupied Europe.

His first stop was to the hospital at Stalag 9C. Here he was treated by a medical staff comprised of volunteers from countries other than Germany who had been brought together by the International Red Cross.

From here he traveled by train to a camp for airmen, Stalag Luft 6, near the town of Heydekrug, Lithuania, where he spent considerable time. Later, the camp was abandoned and the inmates and their guards moved by ship through the Baltic Sea to Stettin (Szczecin), Germany. A new camp was established at this site for a brief period of time.

During the next move, the prisoners were shackled together in pairs, packed into box cars, and traveled by train to the north some 68 miles to a new camp near Swinemünde (Swinoujscie), a port city on the Baltic. Upon arrival, the flyers were unloaded, and while still chained together they were forced to double-time carrying their belongings. German guards with fixed bayonets and guard dogs goaded the prisoners along the route to their new home. During this march, civilians lined the roadsides waiting for the tired and weaker Allied airmen to discard some or all of the their belongings they were carrying. If an exhausted prisoner fell, the one he was shackled to was prodded to help his buddy back to his feet, and the journey continued.

After a short stay in this camp, the prisoners then moved out on a forced march in early February, 1945 for approximately 70 days. They generally headed west, but there was no destination – just keep ahead of the advancing Russian armies. The men marched from dawn until dusk and rested overnight in the cold and chilly fields. Now and then they found a barn to use as shelter for the night.

On 12 April 1945, the 32nd president and Commander in Chief of the Armed Forces of the United States, Franklin Delano Roosevelt died. By mere coincidence on that same day, Sergeant Gudiatis and two other Americans decided to make a break for freedom.

The three GIs slipped away from their large group after dark and headed west. They traveled just the opposite from the method the Germans had been using – the trio slept during the day and moved at night.

As the sun was about to rise on the fifth day of their journey, they again began looking for a hiding place to use during the daylight hours. In their search for a concealed spot, the Americans came upon two Pomeranians who were about to start plowing a field. Gudiatis and his friends approached the workers and attempted to communicate with them. After spending considerable time trying to express themselves, the Americans found out one of the men was Polish while the other was a Russian forced laborer. The two farmers understood the flyers predicament, were sympathetic, and showed the airmen a place where they could conceal themselves.

The three flyers were guests of the Europeans for about a week. Each morning their hosts sent bread, hot tea, or warm potatoes to the tired and hungry GIs.

On the seventh day, the farmers stopped a reconnaissance unit from a British armored division which was traveling on the dirt road that ran by the farm. The British picked up the Americans and took them back to their division rear headquarters for identification, examination, delousing, and debriefing.

From there, the Americans were driven to an airfield near Celle, Germany, located about 19 miles northeast of Hannover, and then flown to England. After arriving in England, they spent time in a field hospital being treated for malnutrition.

Shortly after VE Day, 8 May 1945, John Gudiatis sailed from Southampton, England for the United States and home.

Staff Sergeant Charles J. Groeninger[3]

Lieutenant Lang's left waist gunner, Charles Groeninger, landed on his feet after leaving the aircraft and was uninjured. As he stood talking to a busload of Dutch people and attempting to get out of his parachute harness, he was captured by the German authorities.

While confined in Stalag 17B, Groeninger made an escape attempt with several other inmates on Mothers' Day, Sunday, 14 May 1944.

The prisoners in this camp were always digging escape tunnels. They carried the sand and dirt outside from their burrowing in their pockets or in small cloth bandoliers. Once in the exercise area, the Americans would dispose of their digs by simply shaking them out of their clothes as they walked about.

During World War I, Stalag 17B was used by the Germans as a prisoner of war camp for captured Serbians. They, too, liked to dig, as the GIs soon found out. On several occasions while doing their own digging, the Americans broke through walls and into the tunnels carved out of the ground by the Serbs a quarter of a century earlier. Unfortunately however, the burrows led nowhere.

The American escape plan was simple. First, dig a tunnel out past the three wire fences that surrounded the camp and into the oat field which was adjacent to the wire, second, dig up into the oat field and escape! Thus, the GIs began their tunnel.

In early May, the oats stood close to three feet high. A final check of the tunnel length was made by estimating the distance from the tunnel entrance to the wire fences. The first fence consisted of single strand, knee high warning wire. From this wire it was 12 feet to the second fence which was 10 feet high and covered with barbed wire. Six feet farther out was fence #3. It also was ten feet high and it too was covered with barbed wire. Just beyond the last fence was the guard path constantly in use by the back and forth movement of the sentries.

A meeting was conducted, and a consensus was finally reached. It was determined the tunnel was outside the fence and extended well into the oat field.

Early on Mothers' Day morning preparations were made to break out of the camp at 2000 hours that evening. Later in the morning it began to rain, and continued throughout the day into the evening.

Shortly after 2000 hours Charles Groeninger together with another potential escapee arrived at the end of the tunnel, and the two of them began to burrow upwards to break through the crust of earth above them. After some effort, they broke through the top and found themselves outside the wire *but on the guard path*. Figuring it was too risky to continue, they decided against trying to escape and now worked their way back to the tunnel entrance.

Those waiting to follow were quickly appraised of the situation. Immediately, two more Americans were dispatched to the tunnel's extremity for a second opinion. When they arrived, the point man pushed his head up through the newly created hole and saw the highway to freedom was about four feet short of reaching the edge of the oat field. He also observed a sentry coming his way, but apparently the German soldier had not yet seen him. The second team also backtracked to their comrades and gave them the bad news. Not only could they not escape, but due to the daylong rain, it was impossible to repair and patch the wet ground around the hole on the guard path. The Americans held another meeting and another agreement was reached – the escape attempt was now called off.

Early the next morning, a guard walking his post came across the hole. Immediately a detail of GIs was formed by the Germans, and the airmen were put to work breaking open the entire tunnel then filling it in with dirt.

The flyers were disappointed, but when they held their next meeting the consensus was, "nothing ventured, nothing gained"! Shortly thereafter the Americans began another tunnel.

Chapter 21

The Evaders

Technical Sergeant Thelma B. Wiggins, Jr.[1]

Sergeant Wiggins, the radio operator on the Murdock crew, had been wounded in the right heel by shell fragments just prior to bailing out. During his descent, he drifted away from his crew mates and landed alongside a canal nearly 16 miles southwest of where the remainder of the crew came down.

He settled to the ground near a hamlet called Tongeren which was well inside the Belgian border. He did not realize it at the time but landing this far into Belgium was just what he needed.

As soon as he hit the ground, a man ran up and took his parachute, and as soon as Wiggens took off his Mae West, a woman grabbed it. Speaking in broken English both the man and woman told him to run and hide. He immediately scrambled down the bank of the canal, found a hole among some briers, and hid there.

After several hours of waiting, he noticed two men on bicycles apparently looking for someone. He continued to watch them for several minutes, finally figured they were friendly civilians, and stood up and motioned to them. The men signaled him to lie low, and they came over to his hiding place. They told him to stay put until dark, at which time they would return for him. Later that evening he was moved to a farm house where a nurse treated his wound.

From there, friendly Belgian civilians arranged for his successful escape and evasion through the underground. He arrived in Spain on 29 December 1943 and crossed into Gibraltar on 11 January 1944. He was then flown to London on 17 January 1944, and he subsequently returned to the United States.

Staff Sergeant Lloyd G. Wilson[2]

When Sergeant Wilson, the tail gunner of the Eakle crew who was reluctant to bail out, finally pulled the rip cord of his parachute, he was just high enough above the ground for the chute to barely uncover. After the initial jolt of the opening, Wilson and his partially opened chute plunged into a wooded area. His parachute caught in the tall trees, and he found himself alive and dangling high above the ground.

He was uninjured from the fall, but he was hung up in his harness and could not release himself. Soon numerous people from the nearby village of Eisden, Belgium arrived on the scene, surrounded the tree, and stared up at him.

Someone yelled up to him in broken English, "Are you American or British?"

"American," he shouted back!

He called down and asked the group of Belgians if anyone had a knife which he could use to cut himself loose. No one seemed to understand, so he went through a series of charades indicating just what he wanted to do. They understood this and sent a young boy up the tree who freed him. Wilson then climbed down the tree and looked at his watch. It was just 1350 hours.

He and the small crowd of people walked out of the woods to a clearing adjacent to a dirt road. As he stood pondering his next move, a young girl advanced from the crowd and shyly handed him a tin cup of milk. He thanked her, drank some of the warm liquid, and in return gave her a chocolate bar and some chewing gum. He tried to find out from the girl just where he was, but she was nervous and bashful and would not answer him. He finished the milk and handed the cup back to her. She nodded, smiled, and then quickly ran back to the group of bystanders. He thought surely he must be in Holland or Belgium, especially since no one present had attempted to capture him.

He now separated himself from the gathering and began walking away from the people, trying to find a place to hide. He moved back into the same wooded area where he had just come down and soon found some thick, tall shrubbery and hid in it.

He sat down and took out his "escape purse." Each American flyer was issued one of these kits prior to every mission, in the event he was shot down. Wilson was delighted to find he had Dutch, Belgian, and French money. Besides the currency, he also found it contained detailed, waterproof cloth maps of Europe, a compass the size of a large pea, and a small, steel file.

As he was looking at the maps, trying to orient himself, he heard some noise and took a look from his concealed area. He observed a woman talking excitedly to a man and pointing toward his hiding place. She seemed angry and unhappy, so he figured he had better move out – on the double!

Again, he was able to find some thick shrubs as he moved deeper into the forest, and this time he noticed the area was filled with large piles of fallen leaves. He was exhausted and decided to lie down and rest, covering himself with leaves.

He began to doze off when he heard someone trampling through the woods. He raised up just a bit from his camouflaged spot and saw two German soldiers. They were wearing dark blue uniforms and black steel helmets and had rifles slung over their shoulders. To make matters worse, both of them were walking directly towards him. Wilson figured they had seen him but kept still. When about ten feet away, the two soldiers stopped and unslung their weapons. He waited, but nothing happened. He moved his head just a bit to see what they were doing. Much to his amazement, the uniformed men had leaned their weapons against a nearby tree, and each had lit a cigarette. They had still not seen him lying there under his blanket of leaves.

Several minutes later a whistle blew, and the two soldiers put out their tobacco, picked up their pieces, and left the area. Several hours later two more soldiers came near the vicinity of his hiding place, but again he remained undetected.

Near sundown, a man dressed in civilian clothes approached Sergeant Wilson's position, hissing, "Pst, pst"!

The American raised his head, and replied, "Pst, pst"!

The man approached him and began speaking Flemish, which Wilson could not understand. However, by once again playing "charades", Wilson figured out the Dutchman was telling him to stay put until around 1900 hours. At that time, the man indicated he would return with some civilian clothes for the American.

The tired flyer waited until well after dark, but the man never showed up. He glanced at his watch, and the luminous dial indicated it was 2200. It was now getting cold and damp, so still dressed in his flying togs, he decided to walk in order to keep warm. As he made his way through the timberland, he realized he was making a lot of noise stepping on dried leaves and twigs. Wilson decided he had better get the hell out of the woods as quickly as he could – which he did.

He could now hear peoples' voices, so he very cautiously made his way towards the noise and eventually came out of the woods onto a dirt road. He waited quietly for several minutes, and the sound of the voices disappeared in the distance. He decided to use his compass, stay on the road, and walk all night.

Much to his dismay, when the sun began to rise the next morning he discovered he had walked in a triangle. About 0700 he heard what sounded like a streetcar and went to investigate the noise. He immediately came upon a tram that had stopped to pick up some coal miners.

As he stood in the road watching, a young boy came up to him and asked him if he was English or American. Wilson replied he was American and asked the boy, in "charade language", if he could find him a cigarette. The boy left and reappeared several minutes later with a roughly rolled smoke. He thanked the boy, and gave him an English Shilling. The boy nodded approval, and left.

While watching the miners go to work, the gunner had noticed a nearby farmhouse further down the road, and he now decided to walk in that direction. As he neared the entrance to the dwelling, he approached a middle aged woman standing in the front yard and asked her for a drink of water. She apparently recognized him from the day before when the crowd had gathered, for the woman quickly grabbed his arm and hustled him into the house.

Once inside, the woman gave him some civilian clothes to wear. He changed into them and gave her all his flying gear. She then fed him his first meal since yesterday's breakfast at Chelveston.

At 1100 hours her husband arrived, gave the airman a bicycle, and motioned for Wilson to follow him. They then both peddled off into the warm Dutch morning.

From there, Staff Sergeant Wilson's journey was arranged by the many people of the friendly underground. He was guided through Belgium, France, and crossed the Pyrenees mountains into Spain on 14 January 1944.

He arrived in Madrid the next day, and spent a week under the protection of the United States Military Attache. On 22 January 1944, he crossed into Gibraltar, and two days later he was flown to an airfield in Farnborough, England for further reassignment to the United States.

Staff Sergeant Howard J. Keenan[3]

The right waist gunner for the Lang crew, Howard Keenan, owed a lot to his crew's radio operator, Technical Sergeant Warren E. McConnell, and the engineer, Staff Sergeant Reuben B. Almquist. Just prior to some of the crew bailing out from the waist section of their disabled B-17, McConnell had revived Keenan after he had passed out from lack of oxygen. In addition, Keenan had his chest pack on upside down. Almquist corrected this, then led the groggy gunner to the waist exit, and pushed him out the door.[4]

After getting to the ground, the airman found he had been fortunate enough to land in Holland among many unknown friends. He quickly disappeared into the Dutch underground's pipeline for downed Allied flyers. On Christmas day, 1943, Keenan resurfaced in Paris, France, which was a pivotal checkpoint along the route south to freedom.

It was here he was reunited with his crew mate, the tail gunner from the Lang crew, Steve Krawczynski. Krawczynski, who was also in the pipeline, had arrived in Paris two days ahead of Keenan.

At a meeting towards the end of December, 1943, plans were made to have Steve Krawczynski, the next in line, continue his journey southward from Paris. Howard Keenan, who was also at this gathering of the underground with Krawczynski, pleaded with the committee to let him go next. He was restless and anxious to get on the move again. He realized it was Steve's turn to continue the trip and asked his crew mate if he would let him have his place in line. Steve agreed and the planning group went along with the change. Keenan departed with his contact early the next morning.

Something went wrong during the day – either Keenan blew his cover, or the Germans became suspicious. Whatever the reason, he was arrested by the Gestapo later that afternoon. He was detained briefly in Paris then shipped to the Barth prison camp located in northern Germany near the Baltic Sea.[5] It was here Keenan spent the remainder of the war.

Staff Sergeant Steve Krawczynski[6]

The tail gunner, Sergeant Krawczynski, was the last member of the Lang crew to jump, and due to his free-fall he was the first one to reach the ground. He landed feet first in a farmer's ploughed field just outside the city of Heerlen, Holland.

As he stood in the field pulling in his parachute, a woman walked up to him and handed him two pears. He thanked her, and she nodded back. Almost immediately, a man arrived on the scene and handed him an apple. Both men smiled approvingly at one another. More people arrived and began giving him turnips. He thought for a minute he must be in Texas, for these were the largest turnips he had ever seen! His arms were now full of fruit and vegetables, and he was being overwhelmed by the Hollanders and their generosity.

He decided he had better get into hiding somewhere, so he smiled and kept saying, "Thank you, thank you, thank you", and began handing back the food to the smiling Dutch men and women!

Three of the men in the crowd came forward and began talking to him in English. Steve asked them if they would give him some civilian clothing and help him. At first they seemed willing. However, they told him several Dutch quislings, local political traitors who collaborated with the Nazis, were on their way to arrest the airman. They explained to the American if they were seen helping him they, too, would be taken into custody. Krawczynski said he understood, gathered his chute, and took off running for a wooded area.

He moved deep into the woods and hid his chute, harness, and Mae West under some heavy brush. He then laid down under some small bushes to conceal himself and began to figure out what his next course of action should be.

He had been in the thicket about ten minutes when a young lad of about 14 came into the woods softly calling, "American, American"!

The gunner came out from his hiding place, met with the youth, and they talked. The boy spoke broken English and told the sergeant to stay put until dark when someone would come and help him. Krawczynski did just that, as he remained hidden for the rest of the day.

Later that evening two men came to his hiding place with food and civilian clothing. After he had eaten and changed clothes, they led him to a farm house where they disposed of his military apparel. The two men then made plans for him to escape and evade (E&E) through the efficient, well organized, and coordinated efforts of the underground.

E&E was a long, complicated, and dangerous method of avoiding the enemy and regaining your freedom. It also required a lot of patience. Since the coast of western occupied Europe was crawling with German military awaiting the Allied invasion, the escape and evasion routes were from Holland to Gibraltar via Belgium, France, and Spain.

Sergeant Krawczynski arrived in Paris, France on his way to Spain two days before Christmas, 1943. It was here he was reunited with his crew mate, Staff Sergeant Howard Keenan. It was a brief reunion as Keenan was captured by Germany's secret-police organization, the Gestapo, just days later.

As downed Allied airmen continued their travels through western Europe towards freedom, intelligence information was constantly passed to them. It was hoped they would reach their destination safely and then pass the facts on when they were being debriefed on friendly soil.

Krawczynski was tutored on the fact the man in charge of repelling the Anglo-American invasion from England, Field Marshall Erwin Rommel, was in Tourcoing, France at Christmas. Tourcoing, some eight miles northeast of Lille, was situated along a critical road network close to the French-Belgium border.

He was also reminded not to forget that one month before, in November, a fighter strip containing 18 Bf 109s located just outside Tourcoing was bombed by the Allies. This one air raid on the field had caused little or no damage. In addition, he was briefed to remember since November, all around this town, the Germans were building pillboxes and reinforcing houses for the defense of the area.

Sergeant Krawczynski left Paris in style, riding in an old European model of the American Chevrolet. After a brief trip into the countryside by auto, it was bicycles and walking most of the remainder of the journey. The underground continued to work him farther into the south of France and then finally turned him over to an alpine guide for the journey across the Pyrenees mountains. This rugged terrain feature, which formed the southern border between France and Spain, was the last main obstacle along the escape and evasion route.

Before beginning the long climb over the mountains, he was joined by three other Americans. The four of them proceeded to follow their French leader on foot through the mountains until they reached a desolate area just short of the French-Spanish border. There the guide gave them their final instructions and then bade them good luck and farewell. As directed, they crossed the border at night by fording a small river.

The next day, 12 January 1944, they were apprehended by Spanish authorities and put in a local jail for several days. The American authorities in Spain were notified and quickly obtained the release of the flyers. While the Spanish government was sympathetic towards the Nazis, it looked the other way when the American Consul came to pick up his countrymen who had found their way into Spain.

The four GIs were released from the bastille on the third day and placed on a bus. From there, they traveled to a hotel in Madrid, where they were met by the American Military Attache.

Steve Krawczynski now had to spend some three weeks in Madrid awaiting his time to move into Gibraltar. During his stay in the Spanish capital, he had several experiences he would long remember.

One morning after a late breakfast at the flyer's hotel he decided, since it was such a nice day, he would take a walk around the city. After several hours of sightseeing, he found a bench in a park and sat to rest for a time. After sitting a while, he also realized he needed to find a men's rest room and relieve himself. He asked an elderly Spaniard, who was sitting near him on the bench, if he would direct him to the nearest men's toilet. The man nodded and pointed to a small building standing near them in the park. Steve thanked the gentleman, walked to the building, and went inside. The lavatory consisted of several stalls for privacy, a small basin for washing one's hands, and a single elongated urinal for handling at least 12 men at one time as they stood side by side. The American took his place at the trough and began to empty his bladder. As he stood there, he began reading the graffiti that covered the wall in front of him. Much of it was in Spanish which he had trouble following. Then his eyes settled on an inscription that brought a big smile to his face.

When Krawczynski was growing up as a youngster on Clinton Street in New Britain, Connecticut, he lived across the street from another boy, John Maercia, who was a year younger than Steve. They were playmates, attended the same schools, and grew up together. Shortly after Pearl Harbor, they both enlisted in the Army Air Forces. Later both ended up in England, at the same period of time, as gunners on B-17s but assigned to different bomb groups. They located each other through the mail and agreed to meet in London sometime around Thanksgiving, 1943.

Steve continued to stand in front of the urinal, reading the same handwriting, over and over again, and maintaining a broad smile. The American finished, zipped up, and reread the message to himself one more time, "John Maercia, 455 Clinton Street, New Britain, Connecticut, was here on 5 January 1944!"

Several days later Sergeant Krawczynski had another interesting experience. He, together with some other American airmen, had just returned to the hotel lobby from an afternoon coffee break at a nearby cafe when they were accosted by two Spaniards in their late twenties. They each pulled a pistol on the Americans, and were threatening to shoot them, for they suspected the airmen of being Nazis. There were several heated, verbal exchanges going on between the gunmen and the Americans, and the hotel manager decided to pick up the telephone and called the police.

Finally Steve was able to get one of the gunman aside and began showing him his identification. The other flyers also produced their ID cards, and eventually the Spaniards seemed convinced the Americans were telling the truth. The two vigilantes put away their weapons, apologized for their mistake, shook hands with the airmen, and departed. But that was not the end of the confrontation.

Two police officers arrived several minutes after the gunmen had departed and began interrogating the Americans. The GIs gave the officers a description of the men, and the police insisted the Americans follow them on a tour of the area to try to locate and identify the pair of hoodlums. The Americans diplomatically tried to decline to accompany the police on their search. As far as they were concerned, the matter was over. The police continued to insist, and when it appeared the Spanish "host" authorities were losing their patience, Steve Krawczynski agreed to join the police officers in their search for the two renegades.

The two Spanish policemen together with the American went out the hotel entrance and proceeded to enter each doorway along the street. Proprietors of many shops were questioned and Steve was encouraged to look thoroughly through each store. This procedure continued, with no success, until they reached the end of the block. Here was located what appeared to be a small, two story boarding house or pension. The three searchers stopped outside the entrance to the building, and the police officers began a heated discussion. It appeared to the American the two Spanish authorities could not agree on whether they should enter this building or not. After several more minutes of discussion and arm waving, both police officers nodded agreement, took Krawczynski's arms, and escorted him into the building.

Once inside, the two Spaniards approached the hotel desk and began talking with the manager as the American looked on. This supervisor was a rather short, plump female person wearing heavy make up and clothed in a shear dressing gown. As Steve Krawczynski continued to wait, he gazed about the small lobby of the hotel and noticed the chairs and couches were filled with extremely attractive young ladies. They all smiled at him, and he smiled back. Soon one of the women rose, and approached the GI. She, too, wore a see-through gown, with very little underneath. As she walked by, she smiled and spoke to him in Spanish. He did not understand her Spanish, but he got the distinct impression this place was not a pension.

The two officers now concluded their conversation with the madam, turned, and beckoned for the American to follow them. The three men then began a room by room search of the guest house looking for the two gunmen. This was not a courtesy visit by the police, therefore a polite knock on the door before entering was not in order today. The officers simply threw open the entrances to each of the rooms and with pistols drawn scared the living daylights out of numerous couples, who were in various stages of undress, exercise, etc. The American would then peer into the room and attempt to identify the male as one of the two desperadoes. The search ended when the last room was inspected, and no gunmen were found.

The three men now returned to the front desk. The police officers thanked the "manager", moved to the lobby, bowed to the many ladies in waiting, and arranged to leave. Steve was then escorted back to his hotel. The officers thanked the American and left.

Sergeant Krawczynski arrived in Gibraltar in late January, 1944, was debriefed, then on 6 February 1944 he was flown back to England. After a short stay in the United Kingdom, he was flown by C-54 directly to LaGuardia airport in New York City.

Lieutenant Stanley Alukonis[7]

Like so many others that day Lieutenant Alukonis, Lang's copilot, did not have time to grab a portable bottle of oxygen prior to leaving the aircraft. When he cleared his ship, he immediately deployed his parachute and began his descent from about 20,000 feet. When his chute popped, he received the opening jolt which knocked the breath out of him. This coupled with trying to breathe oxygen from a rarefied atmosphere caused the aviator to begin to suffer from anoxia. He quickly glanced at his watch and noted the time – 1400 hours. Then he tried to light a cigarette, but the match would not burn, only smoke. Again there was not enough oxygen present at this altitude to help ignite the fire. Several Bf 109s made passes at him but did not fire. One came within 15 feet, and he could clearly see the fighter pilot's face.

While still quite high in the air, Alukonis could clearly see the city of Aachen some 15 miles due south of him. He continued to drift northward and descend towards what appeared to be a settlement. He later learned it was Puth, Holland only three miles west of the German border. He, as Keenan and Krawczynski had already done, would soon meet the right people.

Just prior to landing, he realized he was about to settle on top of a barbed wire fence. While trying to maneuver his chute to clear the obstacle, he landed in a pasture and struck the back of his head on the ground. He had missed the fence but was quite dazed from the blow on the head. The Lieutenant sat on the ground for a few minutes, unable to move. From his sitting position, he observed a lot of people running towards him. He thought he was in Germany and asked the first group of persons to arrive if he was in Deutschland.

A women in the gathering replied, "No, Holland!"

He quickly decided it was time to get out of there and hide while he still had the time. He unhooked his parachute harness, and handed it to another woman who was standing there – then he started running onto a dirt road. Two men riding bicycles pulled up along side him as he continued to run. One of them beckoned to him to take his cycle, and Stan hastily accepted. He pedalled several hundred yards, spotted a large ditch, and then abandoned the bike next to a nearby house.

He scrambled down into the ditch and decided he would hide in it for awhile. Figuring no one had seen him enter the gully, he laid there for over an hour.

At approximately 1515 hours, while still hiding in the trench, he saw a four-man German patrol approach his position in their version of the American jeep. Luckily the patrol did not see him and continued on up a dirt road and disappeared into a wooded area.

About and hour and a half before dusk a very old woman came by his hiding place, walked right by him, and dropped a small bottle of milk and several sandwiches on the ground within his reach. The lady gave no indication she knew he was there. She neither broke her stride nor looked right or left but just kept on walking past the ditch until she disappeared in the distance.

He thought to himself, "Well, this is pretty good!"

Just before dark, a young girl came by and sat by the ditch very close to where he was still hiding. She was about 16 or 17, very beautiful, with black hair, black eyes, and perfect features.

Again he thought to himself, "Wow, this isn't bad either!"

She sat there, and he continued to lie perfectly still in the ditch. Neither spoke the other's language so they could not converse. However, he did reach up and hand over his "dog tags" to her hoping this would be a gesture of friendship.

A short time later, a young man of about 18 years of age came by and sat down by the comely damsel. He told the American, in broken English, help was coming. The young couple continued to sit and talk beside the hidden flyer until well after dark.

About 2200 hours two middle aged men approached the ditch. They motioned to Lieutenant Alukonis the three of them should head for cover in a nearby wooded area. The young man and woman left and the American followed after the two men. Once in the woods, one of the men gave the pilot an overcoat and hat to wear to protect him from the cool evening. The pilot now realized he was still wearing only his flying coveralls, and he was cold.

The two Hollanders indicated he should follow them. The three of them then proceeded through the night, following the fence and property lines of the farming community. Fortunately they did not encounter anyone, but the local dogs created a lot of noise, as they barked constantly along the way.

Finally, they came to the little town of Schinnen, Holland where he was taken inside a very modest house. Here he met a third man, and from the conversation the American understood one of the gentlemen was the postmaster of Schinnen. Before turning in, he was shown a small trap door in the middle of the living room floor, which was covered by a rug. Underneath the door was a sandy hole where, in an emergency, a person could jump in and hide. Lieutenant Alukonis spent his first two nights in Holland sleeping on a couch in the living room of this small house.

On the third day, a different man came by the dwelling very early in the morning. He motioned Stan outside and indicated it was time to move on. The Dutchman pointed to one of two bicycles, and indicated the flyer should follow him.

The community was beginning to stir, as many people were going to work. The weather was misty and visibility was only about 400-500 feet which provided excellent cover for the American and his guide. Still wearing his flying coveralls, he shed the overcoat he had received the first night and hopped aboard his cycle. The man climbed onto his bike and took off with Stan following. The man got ahead of the American and continued to pull away from him. Not realizing what was happening, and not wanting to be left behind, Stan tried to catch up with him – but to no avail! There was just no way he could close the distance on the fellow. The faster the American pedalled, the more the Dutchman accelerated. He completely ignored the flyer and would have nothing to do with him – other than keep him in sight. This went on for 12-15 miles with the pilot going all out to keep up with his guide.

The American finally caught up when the Hollander stopped at a very large horse breeding farm. Both of them were completely exhausted from the furious pace they had been keeping. The two men entered the farm house, and the pilot was introduced to another man who apparently owned the farm. The lieutenant was told he would spend the next several days at this location. As the gentleman who led Stan to the farm house was about to depart, Alukonis asked him why he had pedalled away so fast, never allowing him to catch up. The Dutchman smiled and apologized. He then explained in broken English what had happened. During daylight hours, it was imperative the Dutchman keep his distance from the American. He could not allow himself to be caught while associating with an escaping Allied flyer, otherwise the entire escape plan would be compromised. Thus, he made sure there was always plenty of distance between himself and the airman throughout their journey.

During his stay at the farm, the American had quite a few visitors from the Dutch underground. Mainly they came to verify if he was truly an Allied flyer or a German agent trying to infiltrate the escape movement. A number of people came to observe him, ask a few questions, then left. However, it was soon apparent to Alukonis, this is as far as he would go in the underground until those in charge were sure he was who he said he was.

His most thorough interrogation was conducted by a Catholic priest. The Father spoke fluent English and asked many questions concerning the B-17, Stan's life in the United States, and reconfirmed the flyer's serial number. Another priest joined them, and the questioning continued. Eventually the two clergymen seemed convinced the American was who he claimed to be. They now indicated to him a car would pick him up later that evening. The pilot was surprised, as cars were so scarce during the war, but he remained optimistic.

About midnight, a 1937 Ford V-8 (European model) arrived with two men in it. These civilians, both loyal Dutch Nationals, worked for the Germans and so had access to the automobile.

The American was then driven southwest about 16 miles to Maastricht, Holland. It was still dark when the car stopped in front of a home in the middle of the city, and it was here he was hustled into the dwelling of an elderly woman who lived by herself. She was a linguist speaking not only English, French and Flemish but also fluent Polish. Her house was quite large, well partitioned off, and contained numerous rooms located on three floors. At this time, and unbeknown to Alukonis, there were several other Allied flyers staying at this large private residence.

He spent several days here while plans were being made to move him across the Dutch border into Belgium. Each morning he would bid farewell to the lady of the house who would go out on her bike for most of the day. She would return each evening, just prior to dusk, with food for him to eat.

During his stay at the woman's house in Maastricht, Stan spent many of his daylight hours looking skyward while lying on the roof of her brick building. There was a door on the third floor that led to a stairway going up to a skylight in the roof. On numerous occasions, he would climb the stairs, crawl through the opening at the top, and lie on the roof. During this time, he was completely concealed from the street. The weather was good most of the time and besides getting some fresh air, he could look up and watch numerous air battles taking place, as the German fighters harassed and fought the Allied aircraft coming and going to their targets.

The night before he was to leave for Belgium, Stan met another American pilot who was also staying in the house but on a different floor. He was from California and had been shot down on the first Schweinfurt raid on 17 August 1943. This pilot felt it was too dangerous to try to get out and escape, so he was planning to stay in the house and take his chances of not getting caught. Alukonis disagreed with him and said he was going whenever he was given the word to "move out."

Maastricht, Holland was directly on the border with Belgium, and the crossing was well planned and coordinated by the people of both countries involved with the underground. The Dutch plan was to move the Allied airmen, generally one at a time, across the Belgian border and into the large city of Brussels. From here, the Belgian patriots took over.

The word came, and Stan started very early in the morning with several male escorts. Their first obstacle was to cross a foot bridge spanning a canal which in turn crossed over the border. As they made their journey that day, many unknown friends were on the lookout for the three men – letting them know when to stop and when to go.

As they entered Belgium, the American and his companions had to cross many railroad tracks but luckily they saw no one. Shortly after crossing the border, he was taken to a house and spent a day there. It was now 21 October 1943, and Alukonis had been on the run for seven days.

Early the next morning he, together with two other men, left the house. As the trio kept walking, they came to more railroad tracks which they continued to follow. This in turn led them to a train which they boarded. After seating themselves, one of the pilot's friends leaned over and told him when on a train always pretend to doze and make believe he was sleeping. In this way, people would be more likely to leave him alone and not try to engage the fugitive in conversation.

The American was keenly aware of his situation and made every attempt to maintain a low profile. While dressed in civilian clothing, which had been provided by the resistance, he was completely void of any identification. He had no papers, no passport, nothing. His pockets were bare except for a few francs he was carrying which had been provided by his helpers. If he had ever been stopped by the authorities, it would have been the end of the line for him. All he could do was to rely on and place his trust in those who were assisting him – which he did.

After a short ride, the three commuters left the train before it arrived in Brussels. They then boarded a tram that took them into downtown Brussels. They got off the streetcar, walked several blocks, and met another man who was about 50 years old. The four of them went to the older man's apartment, where they shared the small flat for two days.

On the third day, the lieutenant was taken to a small hotel in the heart of the city. He was escorted to the top floor – to a penthouse of sorts. It was here he was interviewed by a woman with a third person interpreting. When the questioning was completed, Stan was returned to the man's flat where he had been staying.

Periodically the American flyer reminded his hosts he should get out of the house once in a while and get some exercise. The Belgians agreed with him. Therefore when Stan Alukonis needed to go out and get some fresh air, his sponsors would arrange a "date" for him. He and his lady friend would take strolls, and they were always well chaperoned. Two gentleman walked in front, while two brought up the rear. All four attendants were well armed with pistols. However, this was always an evening activity that took place after dark, when strict blackout conditions were in effect. He never had any idea where he was, as it was most difficult to see, and he never did get a good look at his date – but it was good to get outside!

During these nightly constitutionals, the American often prayed his group would not bump into some inquisitive German authorities. For if they did, he had the gut feeling there would have been one big, hell of a shoot out – and he was always unarmed!

Once again he was moved, this time to another part of metropolitan Brussels. He would now spend the next 10 days staying with a married couple in their middle thirties.

The wife, who did not work outside the home, busied herself as a homemaker, while the husband appeared to be an extremely dependable person and hard worker. He departed each morning for his job while it was still dark. After putting in some 12 hours of work each day, he would return home after sunset bringing with him food: such as cabbage, potatoes, and bread. The American tried to eat as little as possible, as he realized the couple did not have very much for themselves.

Lieutenant Alukonis was given his own small bedroom during his stay with this couple which provided privacy for all concerned. He slept late many mornings, and then during the day he would assist the woman with any chores that needed to be done around the small house.

On the morning of the fourth day, the flyer slept late as usual. He awakened to find his hostess sitting on the edge of his bed smiling down at him. Not being fully awake, he figured she needed him to help her with some jobs around the domicile. She needed him all right, but it had nothing to do with any tasks concerning the home.

He sat up in bed, pulled his blanket up over his shoulders, and wished her a "good morning" in French. She, too, wished him a "good morning" and continued to smile at him. He now noticed the woman was wearing *only* a housecoat that buttoned down the front and the first four or five fasteners were already unbuttoned exposing a lot of exciting cleavage. Still sitting on the bed by Alukonis, the lady now crossed her legs, and Stan was quick to note most of the lower buttons on her garment were missing – exposing a pair of long, slender legs. Being a very perceptive fellow, the American saw she was wearing lipstick for the first time, her brownish-red hair was neatly coiffured, and she was wearing some kind of intoxicating perfume. He suddenly realized she was an extremely attractive housewife.

As the woman continued to smile at him, she explained to the pilot in broken English her desire to teach him French – all about French. He responded by telling her he had studied the French language for two years in high school, and though not fluent in the language, perhaps that was all the French instruction he needed. She went on to explain he needed a woman to teach and explain to him the intricacies of French culture and love – then she began to remove her garment! With that, the American bounced out of bed, threw his arms out in protest, and politely asked her to stop. He quickly dressed, took the woman's hand, then led her out of the bedroom, and into the living room. As they sat opposite each other in separate chairs, he tried in his best French and English to explain to her his feelings on adultery. While he believed it might possibly be an acceptable policy for some people, the flyer spelled out to her, he could in no way handle this breach of his instilled New England morals.

She listened as he continued. He told her she was not only very beautiful but had a great figure and was extremely desirable. He continued talking with the woman and said he did not wish to offend her. Finally, he told her he could not reconcile himself to the fact it was fitting and proper to go to bed with and make love to the wife of the man who was providing him food and a sanctuary from the enemy. With that, he thanked her for her attentiveness. He now stood, and as diplomatically as possible excused himself, then headed for his room.

Fortunately, he had no more confrontations with his buxom and comely landlady during the remainder of his stay. Later that week the underground picked him up and moved him out of Brussels.

From there, the underground kept moving him constantly southwest from place to place in Belgium and ever so closer to the French border. As he continued his journey, Lieutenant Alukonis was continually reminded a war was going on. The Belgian underground was very active with never ending harassment against the uninvited German occupiers. He could hear the constant explosions of the well planned detonations of installations, the rattle of gunfire from guerrilla attacks, and "his friends" talked a great deal about the many German trains being derailed.

Alukonis' next stop was Tourcoing, France, a small town located just inside the French-Belgium border. It was an easy border crossing. He and his escorts waited for just the right time at night and crossed into France undetected. He was given a billet with a former French army officer who had refused to capitulate to the Germans and had joined the underground.

The American was now in his fourth week of evading the Germans. He spent about six days in Tourcoing and now was becoming very restless. Things were not moving fast enough for Stan, but he did not realize this was the way the underground had to work – deliberate and unhurried.

The impatient American flyer now began making his own plans for a quick escape from the continent. There was a German fighter airfield nearby containing a dozen or so single engine Bf 109s. He asked his supporters to draw him a map of the German airfield, and also requested they get him a pair of German mechanic coveralls and cap to go with them. This they did.

To him his plan was a very simple one. First, he needed to locate the end of the prevailing runway and the "run up" area adjacent to it. According to aviation procedures, the German pilot would taxi his plane to this spot, stop, set his brakes, and make his "before take off" check. Then when he was ready, he would call the control tower for permission to take off.

The American lieutenant next planned to don the German mechanic's clothing, hide in a nearby ditch, and wait at this site for a plane to arrive. When the pilot began his ground check, Stan was going to walk over to the plane, point to the underside of the aircraft, and shake his head as if something was wrong.

According to part three of the American's scenario, the German pilot would then slide back the canopy, and look towards the mechanic for further instructions. Alukonis planned to climb up on the wing, disable the pilot with a wrench, pull the German from the cockpit, and take his place. From there, he would fly the fighter directly to England.

He gave the plan a great deal of thought and was confident it would be successful. He discussed it thoroughly with several underground leaders. However, they told him it was extremely too dangerous for him to try it. Moreover they insisted he could not get away with his scheme, and the Frenchmen refused to provide him any more help. The American became furious with his benefactors! He now was determined to make it on his own!

Alukonis cornered one of the movement leaders, and flatly told him, "I'm going on my way! If you people will not help me, I will escape by myself!"

The American didn't realize it, but he had just said the wrong thing to the right person! Those in charge of the underground now decided it was time to have a formal meeting with the disgruntled flyer.

It began as an amicable gathering with Stan and a number of resistance leaders present. Everyone was seated around a large, rectangular wooden table in the home of the French army officer.

The get-together did not last long, and Stan's sponsors did most of the talking. He was informed, in firm no nonsense language, he could not leave. His escape attempt was considered much too dangerous, and the French leaders seated around the table agreed it most likely would not succeed.

He was further told his narrow minded, selfish attitude could jeopardize the entire underground movement that now flourished between Holland, Belgium, and France. They went on to explain, if he were caught, the Germans had many ways to make a prisoner talk. He was also informed everyone who had been helping him since the day he had been shot down could end up as guests of the enemy. That could only mean many, many people would be subjected to the brutal and usual fatal treatment by the Nazis.

The flyer insisted he could successfully make his getaway, and even if he were caught he would never provide the enemy with any information. The meeting now developed into a heated discussion with Stan's companions becoming upset and irritated at him.

Finally, he was flatly told, "You are not leaving! If you go off by yourself, we will take care of you with our own guns! You have gone far enough with this foolish plan of yours. If you get caught, you will compromise everyone who has been assisting you!"

Another of the leaders told the American, "Don't tell me you will not talk if the Germans get ahold of you! We live among them, and we know how they operate!"

The 25 year-old American was now rapidly being convinced these people were really serious about him not leaving on his own. Even Stan's normally soft-spoken French host, who was present, departed from his polite calm tone of voice, and now gave the pilot an ultimatum – do it their way or no way at all!

After this demand, he began to realize how foolish he had been wanting to leave by himself. It now became clear to him he was about to do something that could jeopardize everyone who had befriended him. Lieutenant Alukonis stayed in Tourcoing and did as he was told.

His next stop was Paris. He arrived there two weeks before Christmas. Once again the pilot met with numerous Catholic priests. They questioned him much as the clergymen had done when he had been staying at the Dutch farm weeks before. It was repetitious, but he remained patient, answered their questions, and did as he was told.

Again, every few days he was moved from place to place. Much like the Dutch and Belgian underground, the French, too, were well organized and efficient.

Lieutenant Alukonis was amazed at the numerous civilians that were either directly or indirectly involved in helping Allied flyers get to safety. Many of them went out of their way to help the airmen get rid of their stress and have them feel relaxed.

As an example, one evening the American's friends arranged a dinner date for him with a very lovely and unattached French woman about his age. He wore his best and only suit which was a conservative dark brown one with a black pin stripe. It did not fit too well. Stan was six feet, two inches tall, and the cuffs of the pants broke several inches above his shoes.

While there was no romantic interest, he and the lady nevertheless enjoyed each other's company, as they dined with candlelight and wine at a nearby cafe. Later that evening, four German officers entered the now crowded restaurant and unbeknown to Stan were seated at the table directly behind him. During the meal, the officer sitting behind Alukonis accidentally backed his chair into the one occupied by the American. The German officer turned around to the American lieutenant, and stared at him – then he smiled and politely said something to him in German. Stan, trying not to choke on a piece of French bread, returned the smile, nodded his head, and turned back to his table. He looked across the table at his date, and she smiled at him as though nothing had happened. However, even in the shadows of the dim, candlelit room, he could make out small beads of perspiration about her forehead. He began to break out in a cold sweat, as he again realized he was not carrying any identification. He and she both took a deep breath, smiled at one and other, raised their wine glasses in a silent toast, and continued their meal.

On another occasion, several civilian men arranged to get him out during the day for some air and exercise. He was given some old clothes and a cap to wear and joined one of their comrades who was pushing a small fish wagon around Paris. The vendor was delighted to have a helper, and he walked while Stan pushed. Their travels carried them to a large fish house adjacent to the Seine river. The Frenchman proudly informed the American the building was over 500 years old. Stan did not doubt it. The stench emitting from the place indicated the first fish ever caught and brought there for sale must still be inside. While inside he had trouble breathing, but the French went about their business as usual.

The next day the American pilot was introduced to a Royal Air Force (RAF) pilot, Ian Campbell, who had recently been shot down at night while flying a British long range, four-engine bomber called a Lancaster. Campbell was also in the underground's pipeline. From here on, the two flyers would remain together as they continued their attempt to successfully escape and evade the enemy.

The two pilots moved several times while staying in Paris. At one stop, they were housed several days with a charming 68-year-old French lady who loved to play bridge. She was delighted to learn both the Englishman and the American played the game, and she insisted the underground provide a "fourth." They complied by

The British Lancaster, a four-engine bomber of the Royal Air Force, similar to the one flown by the English pilot Ian Campbell. (Photo courtesy of Dale E. Remaly)

dispatching a middle-aged French-Canadian woman, whose husband was extremely active in the resistance, to the card table. Thus, in early winter, 1943, an international bridge tournament began in Paris without the Germans being invited.

The tournament ended Christmas Eve, 1943. The two Allied flyers said farewell to their hostess and her friend and during the afternoon moved to a small apartment occupied by two Frenchmen.

As usual the pilots were escorted to their new quarters by several body guards. After their attendants had a short conversation with the new landlords, both airmen were introduced to their hosts, Jean and Pierre. The escort departed, and the flyers were invited inside and told to make themselves comfortable.

As soon as Stan entered the apartment, he realized there was something different about the place. As both pilots sat down on a love seat, the American looked over at Ian, but the Englishman seemed unconcerned. The air in the one room apartment contained a strong fragrance of perfume. One of their hosts, Pierre, was wearing lipstick and had extremely long hair tied back behind his head in a ponytail. Stan glanced again at his ally, but the Englishman still remained complacent.

Pierre pulled back a curtain that separated the kitchen from the living area and began to start the evening meal. Ian now got up from the couch, went to the kitchen area, and asked Pierre if he could help in any way. The Frenchman smiled at the Englishman, said he did not need any help, and told him to rest himself. The cook then blew the Englishman a kiss and moved back to the stove.

Stan had been watching all this, and was now convinced these two Frenchmen deviated from the accepted norm to which he was accustomed. He was also convinced this was going to be a short stay at this place! The English pilot was naive, having been raised by an extremely well-to-do English family, which placed strong emphasis on a heterosexual way of life. Ian was extremely deficient in the sexual behavior of the male species. For example, he still did not realize he and Stan were the guests of two gay Frenchmen.

That evening the four of them celebrated Christmas Eve with a fine bottle of French wine provided by one of the hosts. It was now time to turn in and since there was but one bed Stan, fully clothed, laid down on an old rug that partially covered the floor. Jean, also clothed, laid down beside him and quickly fell sound asleep.

Ian, still dressed, crawled into the lone bed in the flat followed by Pierre. The lights were turned out and just before Stan dozed off he heard a lot of scuffling and noise coming from the other side of the room. He fell asleep sure his English friend could take care of himself.

The next morning, after the two Frenchmen had gone off to work, Stan arose, washed up, and waited for Ian to awake. When the Englishman finally came to, he sat up in bed, and was wild- eyed as he began to talk with his American friend.

"Stan, do you know what that fellow tried to do to me last night?"

The American nodded, and replied, "Yeah, I have a very good idea."

The Englishman was completely shocked at the conduct of the Frenchman and was for leaving the apartment and going on their own. Stan managed to quiet him down and had a better idea.

He and Ian left the flat and made contact with the underground through the woman who had hosted them and the recent bridge tournament. Alukonis and Campbell explained to the resistance leaders the outrage that had taken place the previous evening. Both pilots were informed they must return to their quarters, as they could not be moved again for several days. The flyers were also told they would not be hassled again. They did as they were told, returned to their place, and spent two more days of normal routine with the Frenchmen. The two French hosts had gotten the word, and their guests were not bothered again.

The two Allies were next moved to Bordeaux, France as the underground worked them closer to Spain. It was now 1 January 1944, and they were being moved from place to place almost daily.

Their means of locomotion varied as they went by train, bicycle, boat, and, of course, hiking. All along the way, the two aviators had guides who would take them just so far. Then another scout would pick them up and the three of them would continue the journey.

On 3 January 1944, they arrived at the foothills of the Pyrenees mountains which separated France from Spain. Now came the most strenuous part of their long journey – climbing over the mountains. As the two friends and their guide began the long, arduous climb to the top, they could see snow on some of the peaks. The mountains were a tiring obstacle. After finally ascending to the top of the one they had just finished climbing, they would rest briefly and look ahead at their next barrier. It seemed as there was no end to the high terrain.

The group traveled day and night and rested in sheep pens and any shelters they could find along the way. To further complicate matters, both flyers were wearing low quarter shoes which definitely were not designed for mountain climbing. This, added to the cold weather, snow on the ground, and their constantly wet feet, made the journey even more miserable.

The Englishman, Ian Campbell, looked like a healthy, rosy cheeked young man, but as it turned out he was short on endurance. When they began their upward travel, Ian soon had trouble keeping up with Stan and the guide. He tired quickly and lagged behind much of the time. When the three of them came to a small stream, Campbell always stopped, dropped to his knees, and began drinking the water. This began to cause considerable delays, greatly irritating the advisor, who was trying to maintain a steady pace from point to point along the trail. To encourage the Englander to increase his pace, the scout began whacking him across the rump with the large walking staff he was carrying – always making sure the American was looking the other way when the stick fell. Ian finally complained to Stan about the abuse he was taking, so the American began to help his comrade. The Yank was also toting a walking stick, and he placed it across the back of his shoulders letting Ian hang on as they climbed. This solved the problem.

In late morning of 5 January 1944, the trio stopped to rest seating themselves on the dry, top portion of a sheep pen. The guide pointed in the direction they were traveling, and the two pilots could see it appeared they were approaching the crest of the mountain range. After reaching that pinnacle, it would be down hill all the way!

All throughout the trip, they had been following trails and paths close to and paralleling the Bay of Biscay. This was an intentional choice by the underground planners. Once the border crossing was completed, this prearranged course placed successful escapees near the populated coast of Spain rather than in a remote mountain village.

The three of them crossed the border into Spain on a full moonlit night on 6 January 1944. As in so many mountainous, inaccessible, and uninhabited areas around the world, there were no defined border markers on the ground that separated the two countries. The guide just knew they had arrived where France and Spain meet.

Halfway up a bluff they stopped for a few minutes to rest and get their bearings. It was dark, but the moon was still up. The leader pointed to one of several worn routes and indicated it was the one the three of them would follow. Stan took the point with the scout and Ian following some 10-15 yards behind. The path was a narrow one with large trees bordering each side. It was very still and quiet as the American followed the trail. As Alukonis continued up the pathway, two soldiers on patrol stepped out from behind a tree onto the trail about five feet in front of him. At first, neither saw the other. Seconds later, recognition set in! The startled men just stood in the bright moonlight for a moment staring at each other – then Stan wheeled, and took off back down the trail.

He could hear the two troopers calling, "Halta, Halta," but he kept on running!

He said to himself, "Halta my ass, I'm getting out of here!"

Rifle shots were now whining by his head and snapping into the nearby trees, as he raced back down the footpath. He realized he had never run so fast in his life. Though well behind Stan, the guide and Ian heard the shots, and they too turned and fled. The two flyers finally caught up with each other as they reached the bottom of the hill. Their escort was nowhere to be seen, and the soldiers had not followed.

The two Allies were now on their own, and they decided to keep walking. After stumbling along in the moonlight for some 30 minutes, they could hear dogs barking in the distance. They kept moving and soon heard and saw people up ahead of them who were talking and smoking. They both stopped, backtracked, and got away undetected – together this time.

They now picked a trail that led in a different direction and used the moonlight to help them along their way. After several more minutes of walking, the two of them stopped to try to get a fix on the direction they had been moving since their encounter with the soldiers. They agreed they were still in Spain and should continue along the route they had been following.

Once again they began moving and soon spotted an old farmhouse surrounded by open farmland. They could see light coming through several of the windows and figured someone was home. The two of them decided to wait several minutes to see if anyone else was in the area. Finally convinced they were alone, they decided to approach the dwelling and try to talk to anyone who was inside. Neither of them could speak Spanish, but Campbell was fluent in French, so he volunteered to be the spokesman.

They knocked on the door as politely as they could, and it was shortly opened by a tall, heavy set man in his mid-forties. He looked them over carefully and then invited them in. The house was a modest place which had dirt floors and a large fireplace which heated the main room and where the occupants cooked all their meals.

Campbell and the man immediately began a conversation in French which lasted several minutes. Stan and the man's family sat quietly while the two men chatted. Ian was told they were in one of the Basque provinces of the region of northern Spain on the Bay of Biscay. The farmer spoke a little English also, as years before he had worked in California as a sheepherder. He and the Englishman continued to talk, and after several more minutes of conversation both airmen were offered some "most welcomed" food and drink. After the flyers had eaten, the man indicated the two pilots should spend the evening in his home, and then the next morning he would help them reach their destination.

The farmer, his family, Stan and Ian were all up early the next morning. The man told the flyers how to get to San Sebastian, the city and port in northernmost Spain, located on the Bay of Biscay. He then provided each of them with a bicycle, and a Spanish guide arrived to escort them. The two Allies thanked their benefactor, shook hands, and departed.

After several hours of pedalling, the three riders reached San Sebastian, and as instructed they rode their bikes to the front of a large wooden structure. The pilots said good-bye to the Spaniard who had ridden with them and expressed their gratitude for his help. They then entered the building, introduced themselves to the Spanish authorities, and were told to sit and wait in the lobby. They sat, talked, and wondered what would happen to them next.

After about a 45 minute wait, a gentleman approached them and introduced himself. He was from the British Consul's office in San Sebastian, and he had come to pick them up. Both flyers followed the chap to his limousine, and with the Consul's flags flying the two aviators were driven to Madrid.

After reaching the Spanish capital, the two Allies who had been through so much together, shook hands and took a few moments to say their farewells to each other. They then parted company and moved on with their respective attaches. It was 7 January 1944.

On 14 January 1944, Lieutenant Alukonis took a train from Spain to Gibraltar where he was interrogated by the British. On 17 January 1944, he was flown from Gibraltar to England.

His problems should have but did not end when he arrived back in the United Kingdom. After spening 85 days on "temporary duty" in enemy territory and another 11 days processing through Spain and Gibralter trying to find his way back to his unit, he was rudely greeted by the British and American authorities.

He was interrogated again by the British, and the Americans placed him under house arrest for two weeks. For some reason, no one would believe he had been a member of the 305th Bomb Group. Finally Alukonis made a suggestion to his keepers – perhaps there was someone still alive at the 305th that remembered him. He then proceeded to name his original crew members with which he had flown to England from the United States.

Several days later the bombardier from that crew arrived on the scene and was asked to pick out Lieutenant Stanley Alukonis from a lineup of 20 people. It worked. Alukonis now was able to return to the 305th, where he picked up his records, and orders reassigning him back to the United States.

His next stop was London to hitch a ride back home on an aircraft. Since his orders did not specify a priority, he had to wait for "space available."

On many occasions during the next few days, he was comfortably seated, as a passenger, aboard American aircraft that were planning to fly to the United States. However, each time before take off, he was "bumped" from every flight by someone with a priority.

This practice continued and finally after playing this game of futility for several weeks, Lieutenant Alukonis said, "The hell with this!"

He now decided he would take the first open seat on any Army Air Force aircraft that was headed in *any* direction. He did not care where it was going, he just wanted to get moving – and he did.

He finally found "space available" on a C-47 heading south from London. Having already visited a number of European countries in the past few months, as a desperate escapee, the much traveled American pilot was about to begin another long journey. This time he was seated as a relaxed passenger.

Alukonis' first stop was Morocco (Marrakech), a large city in west central Morocco, located in the foothills of the Grand Atlas mountain range. To reach this northwestern African nation, the plane he was aboard had to circumnavigate Spain and Gibraltar where Stan had been just a month before.

From Morocco he went further south crossing Saguia el Hamra, a territory containing the Spanish Sahara which extended practically the entire length of the area. The flight next passed over Mauritania, a French territory, and finally landed at Dakar, Senegal. This large city and port was also a colony of France.

After a two day lay over, Alukonis boarded another Army Air Force transport and was flown southwest over the Atlantic Ocean to Belem. This city, located on the northern coast of Brazil, was not far from the mouth of the Amazon river. Again, another stopover for several days, before departing for Puerto Rico.

After take off, the aircraft flew northwest paralleling the coasts of French, Dutch, and British Guiana. While out over the Caribbean Sea, the pilot had to shut down one engine of the twin-engine C-47, and the flight was diverted to Jamaica due to the emergency.

From there Lieutenant Alukonis finally caught another flight to Puerto Rico. The last leg of the trip took him from San Juan, Puerto Rico to Miami, Florida and finally home!

Chapter 22

The Turning Point

Shortly after the completion of the 14 October mission, a thorough assessment of what took place was compiled and published by Headquarters, Eighth Bomber Command. This narrative of operations stated 75 percent of the productive capacity of the ball bearing industry at Schweinfurt was lost.[1]

"Strategically it was the most important of the sixteen raids made during the war on the Schweinfurt plants. It caused the most damage and the greatest interference with production, and it led directly to a reorganization of the bearing industry . . . Although the machine damage wrought on 14 October amounted to only 10 per cent and therefore hardly warranted the estimate made shortly thereafter by Allied interpreters that over 50 per cent of Germany's capacity for producing bearings had been destroyed, the damage was felt in critical departments of the industry . . ."[2]

Then on 30 October, a recapitulation of the number of bombers that participated in the raid and what happened to them was issued.[3] This corrected report by Eighth Bomber Command provided the following statistics:

Bomber Command started 377 planes to Schweinfurt, consisting of 317 B-17s and 60 B-24s. Due to mechanical, personnel, and weather problems, the 1st and 3rd Bombardment Divisions had a total of 56 abortions. The sum of 227 B-17s bombed the target. The 2nd Air Division, which dispatched the 60 B-24s, did not complete the mission. The crews encountered poor weather and had difficulty assembling. The "2nd" did manage to gather 48 of its aircraft. However, the division leader elected to fly a diversionary mission, rather than risk going it alone into Germany missing 20 per cent of his strength. This division rendezvoused with 56 P-47s of the 352nd Fighter Group, and took a northerly course up along the Dutch coast to the vicinity of the Frisian Islands. Later, the B-24s returned to their bases in England. No incidents were reported.

The enemy destroyed a total of 60 Eighth Bomber Command B-17 bombers. Thirty-four of them were shot down going to the target while 26 were lost as they fought their way home.

The 1st Air Division put up 163 planes, had 33 aborted missions, lost 29 bombers to the enemy before reaching Schweinfurt, and 101 of its ships managed to bomb the target. The 3rd Bombardment Division sent aloft 154 B-17s, had 23 abortions, lost five ships before arriving over the target, and 126 of its bombers succeeded in unloading their "racks" on Schweinfurt.

The three wings, the 1st, 41st, and 40th of the 1st Division, lost a total of 45 B-17s. This was 75 percent of the entire losses for Eighth Bomber Command.

The 1st Combat Wing, which took over the 1st Division lead shortly after departing the English coast, crossed the enemy "coast in" with 33 B-17s. Its three groups bombed the target with 32 ships. Throughout the entire mission each group lost one ship. One B-17 fell before reaching Schweinfurt, and the other two went down after leaving the target.

The 41st Combat Wing, which flew the number three position all day and brought up the rear of the 1st Division, lost one ship before reaching the target, bombed with 48, and dropped 12 B-17s after coming off the target.

The disorganized 40th Combat Wing, which was directed to head the 1st Bombardment Division but turned over the lead to the 1st Combat Wing over the English Channel and followed it to the target, suffered a crippling blow. This wing lost 27 of the 1st Division's 29 B-17s even before reaching the target. Two more went down after "bombs away"! This loss of 29 ships was from a total of 49 bombers that penetrated the Dutch coast prior to enemy contact. The 305th Bomb Group lost 13 of these aircraft to enemy action, the 306th Group had 10 shot down, and the 92nd Group dropped six. This staggering toll of planes and men amounted to 59 percent of the 40th Combat Wing's assaulting force! It was almost half, 29 out of 60, of the losses suffered by the entire two divisions that participated in the raid!

The 3rd division's losses were just one-third of those of the 1st Division, as 15 of its ships fell to the enemy – five going to the target, and 10 coming home.

Another, five aircraft from the 1st Bombardment Division that returned to England were either lost or severely damaged, as three crews were forced to bail out, and two of them had to crash-land.[4] An additional 63 B-17s from this division received considerable battle damage: 15 of them were classified as receiving major damage while 48 needed minor repairs.[5]

Practically all the 1st Division combat losses took place while the ships were unescorted. The 353rd Fighter Group, which provided friendly P-47 support for this division, reported only three attempts by the German fighters to jump the American B-17s while they were being escorted. The first one, at about 1330, stated about 30 Fw 190s attempted to attack the bombers, but they were driven off by the escorting fighters. The second try, also at 1330, indicated some 20 more enemy fighters made an effort to reach the bombers, but they, too, were turned away. The 353rd Fighter Group told of seeing only one bomber loss prior to the friendly fighters leaving the battle. This was a straggling B-17 that was seen to go down after being attacked by three Fw 190s, who in turn were then attacked by the P-47s.[6]

The B-17 crews of both the divisions claimed 186 German fighters shot down, 27 probably destroyed, plus another 89 damaged by gunfire.[7] Just as with the earlier raid on 17 August 1943, when the Eighth bombed both Schweinfurt and Regensburg and lost 60 bombers, the American total claims for downed enemy aircraft at both these targets were most encouraging. On 17 August 1943, the estimate was 288 destroyed, 37 probably destroyed, and 99 damaged.[8]

Both of the above claims against enemy fighters turned out to be greatly inflated. However, those at the top continued to gaze at these exaggerations through their rose-colored flying glasses. These and other inaccurate figures of enemy aircraft losses from other causes had those in Washington, including General Arnold, looking for an early collapse of the German Air Force.

Fortunately, reality set in. Despite the valiant efforts of the bomber crews and the "careful study of reconnaissance photographs" the destruction of Schweinfurt's ball bearing industry did *not* suffer a 75 percent loss as earlier reported.

As for the number of German fighters destroyed by Eighth Bomber Command gunners, it again became evident that while under heavy attack and with numerous gunners shooting at the same fighter, it was difficult to arrive at an accurate count. After the 17 August mission, enemy losses were recompiled and changed to 25, less than 10 percent of the original estimate of 288.[9] The 14 October figure of 186 destroyed enemy fighters was later pared to 38 – a better figure but only 20 percent of the original number![10]

These revised downward, practical figures of German fighter losses coupled with the increasing destruction of the Eighth's heavy bombers and their crews caused those at the top to finally face the inevitable. Eighth Bomber Command could no longer sustain these incredible losses on the unescorted portion of the deep thrusts into Germany.

For the one month period from 15 September through 14 October, 1943, Eighth Bomber Command had lost some 193 bomber crews out of an estimated monthly available total of 661 – over 29 percent![11]

"The fact was that the Eighth Air Force had for the time being lost air superiority over Germany. And it was obvious that superiority could not be regained until sufficient long-range escort became available. Fighter escort was clearly the answer

The North American P-51 "Mustang" fighter that solved the problem of the long-range bomber escort. (Author's photo)

to the German counterattack, especially to the rocket-firing fighters which, lacking somewhat in mobility, were perculiarly vulnerable to attacks by other fighters...."[12]

The P-51 Mustang, a single-engine fighter, was to solve the problem, but it would not become operational until December, 1943. In March, 1944, the P-51s were fitted with wing tanks that insured they could escort the B-17s and B-24s to any target they chose to bomb.[13]

A belated command decision was finally made. Bomber Command would wait for the arrival of two things: one, the coming of the long-range fighter escort, and two, the arrival of more bombers crews to replace the recent horrendous losses of personnel. So for the remainder of the year, Eighth Bomber Command flew no more deep, unescorted missions into Germany during periods of clear weather.[14]

The wholesale slaughter of the unescorted American bomber crews over Europe had finally come to an end!

Chapter 23

Epilogue

A total of 18 aircraft were dispatched by the 305th to bomb Schweinfurt on 14 October 1943. Three ships aborted and returned early. Another 12 B-17s were lost on the way to the target. Three of the planes bombed the target but one of them, on fire, went down immediately after "bombs away." The remaining two aircraft were able to complete the mission, and return to base.

A total of 13 bombers containing 130 crew members were lost. Of those 130 crewmen, 40 of them died. Of those who survived the mission, 20 were wounded.

Of the 83 who bailed out and lived, 79 of them spent the remainder of the war in various German prisoner of war camps. Four of those individuals who jumped managed to successfully escape and evade the enemy and returned, via the underground, to the United States in early 1944.

Eight flyers of one of the lost crews flew to Switzerland and were interned. However, the crew's navigator died that evening from his wounds.

Upon return from the mission, Major Normand gave his version of what took place. It was obviously not the truth and quite different from what actually happened.

No formal inquiry was requested by Lieutenant Colonel Thomas K. McGehee, the commander of the 305th, or by any of the higher headquarters, as to why the Chelveston outfit had been decimated.[1] Everything that had gone wrong for the aircrews was blamed on an "act of war", caused by the many German fighters that showed up, generally avoided the tight formations, and had attacked the countless stragglers.

In the 1st Bombardment Division's report to Eighth Bomber Command on 20 October 1943, paragraph 3a of the report explained away the 14 October 1943 disaster in one brief sentence. "The serious losses incurred on this operation were occasioned by the vast numbers of fighters the enemy was able to effect . . ."[2]

Instead of being condemned and receiving just punishment for a day of numerous, inept decisions, and poor leadership qualities that greatly contributed to the destruction of his unit's integrity and combat effectiveness, Major Normand's actions were condoned, and he was rewarded for his shortcomings by his superiors.

He was promoted to the rank of Lieutenant Colonel on 27 October 1943, just 13 days after the mission.³ He had been in grade as a Major since 17 February 1943 – some eight months.⁴

Several weeks after the 2nd Schweinfurt raid, Lieutenant Colonel Charles G.Y. Normand was awarded the Silver Star which is presented to American military heroes "For Gallantry in Action."⁵

Normand continued to lead a charmed life for awhile. He kept on flying missions and took his turn leading the group. However, his luck ran out on 24 August 1944. On that day, while on a mission to Leipzig, Germany, he was shot down, captured, and became a prisoner of war in Stalag Luft 3.⁶

In Retrospect

With Major Normand leading, the 305th failed to join the 92nd and 306th Bomb Groups over England, on 14 October 1943, to form the 40th Combat Wing. This created a "no win" situation for Colonel Budd Peaslee, the leader of the 40th and many of the bomber crews from that wing. Peaslee was left with three options:

1. He could have aborted the mission, and ordered the return of the 92nd and 306th groups to their bases. This would have reduced the 1st Division bomber force by 34 aircraft (28%).

2. He could have maintained the division lead with his two groups, and eventually Normand may have caught up with him, as he did chasing Milton's 1st Combat Wing. At that time, had Peaslee known Lieutenant Colonel Milton was also short a group, he might have opted for this choice and retained the lead.

3. He could have turned over the lead to the wing behind him, and attempt to keep his two groups intact as he maneuvered in behind the new leader. He elected to execute this third option and the disastrous 2nd Schweinfurt raid is now history.

For the many American B-17 crew members that fell that day, the war finally ended. The German prisoner of war camps opened the gates and freed their weary, sick, and hungry inmates. Many of those wounded, healed. Psychologically, a number of them would never get well. Most of the dead flyers, interred by their enemies, were located and eventually reburied in their final resting places. Then there were those who could not be found and are remembered on the "Tablets of the Missing."

Appendix A

The two 305th BG aircraft and crews that were able to complete the mission and return to Chelveston

Crew member	Position
A/C #42-3412	
Joseph W. Kane	p
Charles G.Y. Normand	cp*
Jack J. Edwards	nav
Joseph Pellegrini	bomb
LeRoy V. Pikelis	top t
Owen R. Hanson	radio
B.T. Davis	ball
G.M. Roe	rt w
Charles B. Lozenski	lt w
H.W. Luke	tail
A/C #42-30678	
Frederick B. Farrell	p
Roy A. Burton	cp
Max Guber	nav**
Frederick E. Helmick	bomb**
Marvin D. Shaull	top t
Roger J. Goddard	radio
James F. Higdon	ball
Jayson C. Smart	rt w
Luther Bonones	lt w
Thaddeus J. Niemiec	tail

* also aircraft commander
** wounded

Appendix B

Eighth Air Force Aircraft and Crews Lost from the 305th Bombardment Group (H) on 14 October 1943 listed in sequence as they were shot down

1.	B-17F	42-29952	Douglas L. Murdock	364th
2.	B-17F	42-3436	Dennis J. McDarby	364th
3.	B-17F	42-30807	Gerald B. Eakle	364th
4.	B-17G	42-3549	Charles W. Willis, Jr.	366th
5.	B-17F	42-29988	Robert W. Holt	364th
6.	B-17F	42-30831	Edward W. Dienhart	364th
7.	B-17F	42-3195	Verl D. Fisher	366th
8.	B-17G	42-37750	Robert S. Lang	366th
9.	B-17F	42-30242	Ellsworth H. Kenyon	364th
10.	B-17F	42-30814	Robert A. Skerry	366th
11.	B-17F	42-30804	Victor C. Maxwell	365th
12.	B-17G	42-3550	Alden C. Kincaid	365th
13.	B-17G	42-37740	Raymond P. Bullock	365th

A/C #42-29952 (364th)

Crew Member	Psn	Wounded	KIA	Bailed Out and POW
Douglas L. Murdock	p		x	
Edwin L. Smith	cp			x
John C. Manahan	nav		x	
John E. Miller	bomb		x	
Russell J. Kiggins	top t		x	
Thelma B. Wiggens, Jr.*	radio	x		x*
John W. Lloyd	ball			x
Lester J. Levy	rt w			x
Tony E. Dienes	lt w		x	
William B. Menzies	tail			x

*escaped and evaded

A/C #42-3436 (364th)

Crew Member	Psn	Wounded	KIA	Bailed Out and POW
Dennis J. McDarby	p			x

Crew Member	Psn	Wounded	KIA	Bailed Out and POW
Donald P. Breeden	cp		x	
William J. Martin	nav		x	
Harvey A. Manley	bomb		x	
Arthur E. Linrud	top t			x
Hosea F. Crawford	radio			x
Benjamin F. Roberts	ball			x
Robert G. Wells	rt w		x	
Leonard R. Henlin	lt w		x	
Dominic C. Lepore	tail	x		x

A/C #42-30807 (364th)

Crew Member	Psn	Wounded	KIA	Bailed Out and POW
Gerald B. Eakle	p			x
Walter H. Boggs	cp		x	
Charles B. Jones	nav	x		x
Herman E. Molen	bomb	x		x
Barden G. Smith	top t			x
Donald H. Norris	radio			x
Robert L. Sanchez	ball			x
Lloyd F. Knapp, Jr.	rt w			x
Alfredo A. Spadafora	lt w	x		x
LLoyd G. Wilson*	tail			x*

*escaped and evaded

A/C #42-3549 (366th)

Crew Member	Psn	Wounded	KIA	Bailed Out and POW
Charles W. Willis, Jr	p		x	
John J. Emperor	cp		x	
Christian W. Cramer	nav			x
Willard E. Dixon	bomb			x
Floyd J. Karns	top t			x
Alan B. Citron	radio			x
Bernard E. Snow	ball			x
John L. Gudiatis	rt w	x		x
Edward J. Sedinger	lt w		x	
Nicholas Stanchak	tail		x	

A/C #42-29988 (364th)

Crew Member	Psn	Wounded	KIA	Bailed Out and POW
Robert W. Holt	p		x	
Aubrey C. Young, Jr.	cp		x	
Edward O. Ball	nav		x	
Bryce Barrett	bomb		x	
Charles E. Blackwell	top t	x		x
Doris O. Bowman	radio		x	
William C. Frierson	ball			x
Floyd A. Lenning	rt w			x
Frank W. Rollow	lt w			x
Mike S. Letanosky	tail		x	

A/C # 42-30831 (364th)

Crew Member	Psn	Wounded	KIA	Bailed Out and POW
Edward W. Dienhart**	p			
Brunson W. Bolin	cp			x
Donald T. Rowley**	nav		x	
Carl A. Johnson**	bomb	x		
George H. Blalock, Jr.	top t			x
Hurley D. Smith**	radio	x		
Raymond C. Baus**	ball			
Robert Cinibulk**	rt w			
Christy Zullo**	lt w			
Bernard Segal**	tail			

**crashed landed in Switzerland and interned

A/C #42-3195 (366th)

Crew Member	Psn	Wounded	KIA	Bailed Out and POW
Verl D. Fisher	p			x
Clinton A. Bush	cp			x
Carl H. Booth, Jr.	nav		x	
Donald L. Hissom	bomb		x	
Clinton L. Bitton	top t			x
Harvey Bennett	radio	x		x

Crew Member	Psn	Wounded	KIA	Bailed Out and POW
Harold Insdorf	ball		x	
Thomas E. Therrien	rt w		x	
Loren M. Fink	lt w	x		x
George G. LeFebre	tail			x

A/C #42-37750 (366th)

Crew Member	Psn	Wounded	KIA	Bailed Out and POW
Robert S. Lang	p		x	
Stanley Alukonis*	cp			x*
John C. Tew, Jr.	nav			x
James G. Adcox	bomb			x
Reuben B. Almquist	top t			x
Warren E. McConnell	radio			x
Kenneth A. Maynard	ball			x
Howard J. Keenan	rt w			x
Charles J. Groeninger	lt w			x
Steve Krawczynski*	tail			x*

*escaped and evaded

A/C #42-30242 (364th)

Crew Member	Psn	Wounded	KIA	Bailed Out and POW
Ellsworth H. Kenyon	p			x
Thomas H. Davis	cp			x
John A. Cole	nav			x
Joseph F. Collins, Jr.	bomb			x
Finley J. Mercer, Jr.	top t			x
Russell R. Ahlgren	radio		x	
Arthur Englehardt	ball			x
Richard W. Lewis	rt w			x
Charles M. Green	lt w	x		x
Walter L. Gottshall	tail			x

A/C #42-30814 (366th)

Crew Member	Psn	Wounded	KIA	Bailed Out and POW
Robert A. Skerry	p			x

Crew Member	Psn	Wounded	KIA	Bailed Out and POW
John C. Lindquist	cp			x
Robert Guarini	nav	x		x
Cecil S. Key	bomb			x
Robert J. Middleby	top t			x
Stanley H. Larrick	radio		x	
Wayne D. Rowlett	ball			x
Jack G. Johnson	rt w		x	
Gus Doumis	lt w			x
Edwin E. DeVaul	tail	x		x

A/C #42-30804 (365th)

Crew Member	Psn	Wounded	KIA	Bailed Out and POW
Victor C. Maxwell	p		x	
Willis V. Rowan	cp		x	
Urban H. Klister	nav		x	
Andrew J. Zavar	bomb		x	
Silas W. Adamson	top t		x	
Jerome B. Pumo	radio		x	
Craig T. Conley	ball		x	
Charles H. Crane	rt w	x		x
Herbert S. Whitehead	lt w			x
William H. Connelley	tail			x

A/C #42-3550 (365th)

Crew Member	Psn	Wounded	KIA	Bailed Out and POW
Alden C. Kincaid	p	x		x
Norman W. Smith	cp		x	
Robert D. Metcalf	nav			x
Phillip A. Blasig	bomb		x	
John F. Raines	top t			x
Bernard T. Martin	radio		x	
Kenneth R. Fenn	ball	x		x
Alfred C. Chalker	rt w			x
Louis Bridda	lt w			x
William C. Heritage	tail			x

A/C #42-37740 (365th)

Crew Member	Psn	Wounded	KIA	Bailed Out and POW
Raymond P. Bullock	p			x
Homer L. Hocker	cp	x		x
Manning L. Lawrence	nav			x
Walter A. Kaeli	bomb			x
Carl J. Brunswick	top t			x
Harold E. Jackson	radio			x
Joseph K. Kocher	ball			x
Harold E. Coyne	rt w			x
Stanley J. Jarosynski	lt w			x
Alden B. Curtis	tail			x

Appendixes

Appendix C

In Memoriam

Listed below are the names of the officers and noncommissioned officers, by crew and position, who made the supreme sacrifice on 14 October 1943. Also listed is their final resting place.

Murdock Crew
Douglas L. Murdock, pilot, Greenwood Cemetery, Barnesville, Georgia.
John C. Manahan, navigator, Netherlands American Cemetery and Memorial, Margraten, Netherlands.
John E. Miller, togglier, a private cemetery near Santa Barbara, California.
Russell J. Kiggens, top turret gunner, Netherlands American Cemetery and Memorial, Margraten, Netherlands.
Tony E. Dienes, left waist gunner, Netherlands American Cemetery and Memorial, Margraten, Netherlands.

McDarby Crew
Donald P. Breeden, copilot, his remains were never recovered, therefore his name is engraved on the Tablets of the Missing at the Netherlands American Cemetery and Memorial, Margraten, Netherlands.
William J. Martin, navigator, Netherlands American Cemetery and Memorial, Margraten, Netherlands.
Harvey A. Manlcy, bombardier, Netherlands American Cemetery and Memorial, Margraten, Netherlands.
Robert G. Wells, right waist gunner, Pythian Ridge Cemetery, Sturgis, Kentucky.
Leonard R. Henlin, left waist gunner, Netherlands American Cemetery and Memorial, Margraten, Netherlands.

Eakle Crew
Walter H. Boggs, copilot, Ardennes American Cemetery and Memorial, Neupre (Neuville-en-Condroz), Belgium.

Willis Crew
Charles W. Willis, Jr., pilot, Netherlands American Cemetery and Memorial, Margraten, Netherlands.
John J. Emperor, copilot, St. Joseph's Cemetery, Auburn, New York.

Left: The gravesite of John C. Manahan, Murdock crew, located in the Netherlands American Cemetery and Memorial, Margraten, Netherlands. (Photo courtesy Ron W.M.A. Putz, The Netherlands)

Below: The gravesite of Douglas L. Murdock, Murdock crew, located in the Greenwood Cemetery, Barnesville, Georgia. (Author's photo)

Edward J. Sedinger, left waist gunner, Ardennes American Cemetery and Memorial, Neupre (Neuville-en-Condroz), Belgium.
Nicholas Stanchak, tail gunner, East Ridgeland Cemetery, Clifton, New Jersey.

Holt Crew
Robert W. Holt, pilot, Jefferson Barracks National Cemetery, St. Louis, Missouri.
Aubrey C. Young, Jr., copilot, Golden State National Cemetery, San Bruno, California.
Edward O. Ball, navigator, Jefferson Barracks National Cemetery, St. Louis, Missouri.
Bryce Barrett, bombardier, Jefferson Barracks National Cemetery, St. Louis, Missouri.
Doris O. Bowman, radio operator, Jefferson Barracks National Cemetery, St. Louis, Missouri.
Mike S. Letanosky, tail gunner, St. John's Cemetery, Covington, Kentucky.

Dienhart Crew
Donald T. Rowley, navigator, a private cemetery in California.

Fisher Crew
Carl H. Booth, navigator, his remains were never recovered, therefore his name is engraved on the Tablets of the Missing at the Netherlands American Cemetery and Memorial, Margraten, Netherlands.
Donald L. Hissom, togglier, Northview Cemetery, New Martinsville, West Virginia.
Harold Insdorf, ball turret gunner, Bethel Cemetery, Bergen City, New Jersey.
Thomas E. Therrien, right waist gunner, Ardennes American Cemetery and Memorial, Neupre (Neuville-en-Condroz), Belgium.

Kenyon Crew
Russell R. Ahlgren, radio operator, Fairview Cemetery, New Britain, Connecticut.

Lang Crew
Robert S. Lang, pilot, Ardennes American Cemetery and Memorial, Neupre (Neuville-en-Condroz), Belgium.

Skerry Crew
Stanley H. Larrick, radio operator, Martinsville Cemetery, Clinton City, Ohio.
Jack G. Johnson, right waist gunner, Milford Cemetery, Milford, Texas.

Maxwell Crew
Victor C. Maxwell, pilot, Lorraine American Cemetery and Memorial, St. Avold (Moselle), France.
Willis V. Rowan, copilot, somewhere in Florida.

Above: The name of Carl H. Booth, Jr. engraved on the Tablets of the Missing at the Netherlands American Cemetery and Memorial, Margraten, Netherlands. Below: The name of Donald P. Breeden engraved on the Tablets of the Missing at the Netherlands American Cemetery and Memorial, Margraten, Netherlands. (Both photos courtesy of Ron W.M.A. Putz, The Netherlands)

Urban H. Klister, navigator, St. Paul's Cemetery, Wrightstown, Wisconsin.
Andrew J. Zavar, bombardier, Lorraine American Cemetery and Memorial, St. Avold (Moselle), France.
Silas W. Adamson, top turret gunner, in a private cemetery near Indiana, Pennsylvania.
Jerome B. Pumo, radio operator, Lorraine American Cemetery and Memorial, St. Avold (Moselle), France.
Craig T. Conley, ball turret gunner, McCallsburg Cemetery, McCallsburg, Iowa.

Kincaid Crew
Norman W. Smith, copilot, Lorraine American Cemetery and Memorial, St. Avold (Moselle), France.
Philip A. Blasig, bombardier, Rock Island National Cemetery, Rock Island, Illinois.
Bernard T. Martin, radio operator, in a private cemetery near Fairfield, Montana.

Notes

Book I – The Mission
Chapter 1 – The Eighth Air Force

1. Wesley Frank Craven and James Lea Cate, *The Army Air Forces in World War II*, Volume II, (University of Chicago, 1949), pp. 349-354.

2. Ibid., p. 373

3. Ibid., p. 366

4. Combined Operational Planning Committee, *Tactical Plan for the Attack of Schweinfurt, 2 August 1943*, refers to the "P-47s with long-range tanks" and "the long-range P-47s." When using the belly tank to increase range, the P-47 was still only able to reach the German border.

5. Wesley Frank Craven and James Lea Cate, *The Army Air Forces in World War II*, Volume II, (University of Chicago, 1949), pp. 679-680.

6. Ibid., pp. 845-847.

7. Headquarters, Eighth Bomber Command, *Bomber Command Narrative of Operations, Day Operations, 17 August 1943*, Mission No.84, p. 1.

8. Ibid., p. 3.

9. Ibid., pp. 1, 3.

10. Wesley Frank Craven and James Lea Cate, *The Army Air Forces in World War II*, Volume II, (University of Chicago, 1949), p. 848.

11. Ibid., p. 850.

12. USAF Historical Research Agency, *History of the 381st Bombardment Group*, Maxwell AFB, Montgomery, Alabama.

13. Statement from Charles J. Groeninger, Lang crew, 1988.

14. Statement from Richard W. Lewis, Kenyon Crew, 1988.

Chapter 2 – The Aircraft and Its Crew

1. Headquarters, Eighth Bomber Command, *Bomber Command Narrative of Operations, 14 October 1943*, pp. 1-2.

2. Edward Jablonski, *Flying Fortress*, (New York, Doubleday, 1965), pp. 5-6.

3. Ibid., pp. 310-311.

4. Statement from Edward W. Dienhart, Dienhart crew, 1988.

5. Office of the Operations Officers from the 364th, 365th, and 366th Bombardment Squadrons to the Operations Officer of the 305th Bombardment Group (H), *Combat Crews on Mission of 14 October 1943*, dated 14 October 1943. These lists included the aircraft serial numbers which were collated with a list of B-17s issued to the 305th Bomb Group between 1942 and 1943.

6. Edward Jablonski, *Flying Fortress*, (New York, Doubleday, 1965), p. 338.

7. 305th Bomb Group Combat Crew Handbook, Book No. 102, 1 October 1943, revised 2 February 1945, p. 16.

8. Terry A. Knight, Confederate Air Force, provided the technical information from B-17G manuals.

9. For this story, the author measured the interior dimensions of the B-17F, "Memphis Belle", in 1988, which was located at Mud Island, Memphis, Tennessee.

10. 305th Bomb Group Combat Crew Handbook, Book No. 102, 1 October 1943, revised 2 February 1945, p. 34.

11. Terry A. Knight, Confederate Air Force, provided the technical information from B-17G manuals.

12. This comment is based on the author's flying experience, and some B-17 pilots may not agree.

Chapter 3 – The Briefing

1. Office of the Operations Officer, 305th Bombardment Group (H), to Commanding Officer, 305th Bombardment Group (H), *Narrative Report of Operations of the 305th Bombardment Group (H), AAF, Schweinfurt, 14 October 1943*, dated 15 October 1943.

2. Ibid.

3. Statement from William C. Frierson, Holt crew, 1989.

4. Statement from Charles J. Groeninger, Lang crew, 1988.

5. Statements from Kincaid crew members, John F. Raines, 1988, and William C. Heritage, 1988.

6. Robert O'Hearn, *In My Book You're All Heroes*, (California, Hall Letter Shop, 1984, 1988,) pp. 105, 107; and statements from Bullock crew members Brunswick, Kocher, and Jarosynski resolved the following: Raymond P. Bullock, Carl J. Brunswick, Stanley J. Jarosynski, and Alden B. Curtis were all flying their 25th mission. Joseph K. Kocher and Harold E. Coyne were on their 24th mission.

7. Statements from Frederick E. Helmick, Farrell crew, 1989, and Herman E. Molen, Eakle crew, 1988.

8. Togsliers were gunners who had been specifically trained to permanently replace a limited number of squadron bombardiers throughout Eighth Bomber Command. In late fall, 1942, Colonel Curtis E. LeMay, then commander of the 305th, changed the group's bombing tactics. In a letter written to General Olds on 12 January 1943, LeMay explained when the 305th Group began flying missions their wing was bombing by three ship elements. LeMay developed and employed squadron (six ship) bombing tactics utilizing togsliers. When the squadron lead bombardier dropped his bombs, the togsliers flipped a bomb release toggle switch and bombed with him. Using the togglier, bombing as a group came later when all ships would drop on the group lead bombardier.

9. Statement from Lester J. Levy, Murdock crew, 1988.

10. Statement from Edwin L. Smith, Murdock crew, 1988.

11. Statement from Joseph K. Kocher, Bullock crew, 1988.

12. Headquarters, *Eighth Bomber Command*, Operation No. 115, Digest of Field Orders, 14 October 1943, p. 1.

13. Ibid., pp. 1, 4.

14. Ibid., p. 2.

15. Ibid., pp. 2-3.

16. Ibid.

17. Statement from Joseph W. Kane, Kane crew, 1989.

18. Statement from Delmar E. Wilson, Major General, USAF (Ret), 1989.

19. AAF Form No. 5, Individual Flight Records of Charles G.Y. Normand, provided by Flight Records Section, Norton AFB, California.

20. Wesley Frank Craven and James Lea Cate, *The Army Air Forces in World War II*, Volume I, (University of

Chicago, 1949), pp. 526-537.

21. AAF Form No. 5, Individual Flight Records of Charles G.Y. Normand, provided by Flight Records Section, Norton AFB, California.

22. Statement from Delmar E. Wilson, Major General, USAF (Ret), 1989.

23. Ibid.

24. Statement from Thomas K. McGehee, Lieutenant General, USAF (Ret) 1990.

25. AAF Form No. 5, Individual Flight Records of Charles G.Y. Normand, provided by Flight Records Section, Norton AFB, California.

26. Headquarters, 40th Combat Wing, *Supplement No. 1, Inclosure "A" to 1st Bombardment Division Field Order 220*, to the 305th Bombardment Group (H), 14 October 1943, p. 1.

27. Ibid.

28. Eighth Bomber Command Operational Route Forecast for 14 October 1943.

29. This information on the enemy antiaircraft portion of the briefing is based upon a consensus of questionnaires submitted to the author by 39 of the surviving 305th crew members that flew on the 14 October 1943 mission to Schweinfurt.

30. Headquarters, Eighth Bomber Command, *Bomber Command Narrative of Operations, Day Operations, 17 August 1943, Mission #84*, p. 1.

31. Headquarters, 40th Combat Wing, *Supplement No.1, Inclosure "A" to 1st Bombardment Division Field Order 220*, to the 305th Bombardment Group (H), 14 October 1943, p. 1.

32. This information on the enemy fighter portion of the briefing is based upon a consensus of questionnaires submitted to the author by 39 of the surviving 305th crew members that flew on the 14 October 1943 mission to Schweinfurt.

33. Ibid.

34. Headquarters, 40th Combat Wing, *Supplement No.1, Inclosure "A" to 1st Bombardment Division Field Order 220* , to the 305th Bombardment Group (H), 14 October 1943, p. 1.

35. Ibid.

Chapter 4 – Take Off and Assembly

1. Statement from Charles B. Lozenski, Kane crew, 1988.

2. Office of the Operations Officer from the 365th Bombardment Squadron, to the Operations Officer of the 305th Bombardment Group (H), *Combat Crews on the Mission of 14 October 1943*, dated 14 October 1943.

3. Statement from Joseph K. Kocher, Bullock crew, 1988.

4. Statements from Eakle crew members, Herman E. Molen, 1988, and Gerald B. Eakle, 1989.

5. Statement from Alden C. Kincaid, Kincaid crew, 1988.

6. Statement from William C. Heritage, Kincaid crew, 1989.

7. Ibid.

8. Statement from Christy Zullo, Dienhart crew, 1988.

9. Statement from Raymond C. Baus, Dienhart crew, 1988.

10. Statements from Farrell crew members, Frederick E. Helmick, 1988, and Roy A. Burton, 1989.

Notes

11. Statement from Roy A. Burton, Farrell crew, 1989.

12. Statements from Farrell crew members, Roy A. Burton and Jayson C. Smart, 1989.

13. Entries in 305th Bomb Group Take Off Log, 14 October 1943.

14. Ibid.

15. Entry in 305th Bomb Group lead navigator's log, 14 October 1943.

16. Entry in the 305th Bomb Group Take Off Log 14 October 1943.

17. 305th Bomb Group Combat Crew Handbook, Book No. 102, 1 October 1943, revised 2 February 1945, p. 13.

18. Statement from Edwin L. Smith, Murdock crew, 1988.

19. Headquarters, 40th Combat Wing, *Supplement No.1, Inclosure "F" to 1st Bombardment Division Field Order 220*, to the 305th Bombardment Group (H), 14 October 1943.

20. Headquarters, 305th Bombardment Group (H), Office of the Operations Officer, *Description of Navigator's Problem on Mission of 14 October 1943*, to Commanding Officer, 305th Bombardment Group (H), dated 15 October 1943, p. 1.

21. Thomas M. Coffey, *Decision Over Schweinfurt*, (New York, David McKay, 1977), p. 288.

22. Headquarters, *Eighth Bomber Command*, Enemy Tactics, Schweinfurt – Bomber Command Attack, 14 October 1943, p. 1.

23. Entry in the 92d Bomb Group lead navigator's log, 14 October 1943.

24. Ibid.

25. Entry in the 305th Bomb Group lead navigator's log, 14 October 1943.

26. Statement from Harry J. Task, Task crew, 1989.

27. Headquarters, 305th Bombardment Group (H), *Abortive Report for Mission of 14 October 1943*, dated 15 October 1943.

28. Entry in the 305th Bomb Group lead navigator's log, 14 October 1943.

29. Entry in the 92d Bomb Group lead navigator's log, 14 October 1943.

30. Headquarters, 1st Bombardment Division, *Report of Operations*, to the Commanding General, Eighth Bomber Command, 20 October 1943.

31. Martin Caiden, *Black Thursday*, (New York, E.P. Dutton, 1960), p. 132.

32. Headquarters, 305th Bombardment Group (H), Office of the Operations Officer, *Description of Navigator's Problem on Mission of 14 October 1943*, to Commanding Officer, 305th Bombardment Group (H), dated 15 October 1943, p. 1.

33. Entry in the 92d Bomb Group lead navigator's log, 14 October 1943.

34. Ibid.

35. Entry in the 305th Bomb Group lead navigator's log, 14 October 1943.

36. Ibid.

37. Statement from William C. Frierson, Holt crew, 1989.

38. Entry in the 305th Bomb Group lead navigator's log, 14 October 1943.

39. Headquarters, 305th Bombardment Group (H), Office of the Operations Officer, *Description of Navigator's*

Problem on Mission of 14 October 1943, to Commanding Officer, 305th Bombardment Group (H), dated 15 October 1943, p. 1.

40. Entry in the 92d Bomb Group lead navigator's log, 14 October 1943.

41. Entries in the 92d and 305th Bomb Group lead navigators' logs, 14 October 1943.

42. Entries in the 305th Bomb Group lead navigator's log, 14 October 1943.

43. Entries in the 92d Bomb Group lead navigator's log, 14 October 1943.

44. Entries in the 92d and 305th Bomb Group lead navigators' logs, 14 October 1943.

45. Headquarters, 305th Bombardment Group (H), Office of the Operations Officer, *Description of Navigator's Problem on Mission of 14 October 1943*, to Commanding Officer, 305th Bombardment Group (H), dated 15 October 1943, p. 1.

46. Statement from Arthur E. Linrud, McDarby crew, 1989.

47. Statement from William C. Frierson, Holt crew, 1989.

48. Statement from Ellsworth E. Kenyon, Kenyon crew, 1989.

49. Statements from Farrell crew members, Frederick B. Farrell, 1988, and Frederick E. Helmick, 1988.

50. Statement from Jayson C. Smart, Farrell crew, 1989.

51. Statement from Carl J. Brunswick, Bullock crew, 1989.

52. Statement from Joseph K. Kocher, Bullock crew, 1988.

53. Statement from LeRoy V. Pikelis, Kane crew, 1989.

54. Colonel Budd J. Peaslee, "The World's Greatest Air Battle", (*True Magazine*, April, 1957) p. 123.

55. Entries in the 305th Bomb Group lead navigator's log indicated the group departed Orfordness at 1232 and made landfall over the Dutch coast at 1304 hours.

Chapter 5 – Across the Channel

1. Headquarters, Eighth Bomber Command, *Inclosure "F" to Operation No. 115, Digest of Field Orders*, 14 October 1943.

2. Statement from William E. Lutz, Reid crew, 1989.

3. Headquarters, 305th Bombardment Group (H), *Abortive Report for Mission of 14 October 1943*, dated 15 October 1943.

4. Entry in 305th Bomb Group lead navigator's log, 14 October 1943.

5. Statement from Ellsworth E. Kenyon, Kenyon crew, 1989.

6. Life jackets used by American flyers were affectionately called a "Mae West" by the servicemen. Each of the two sides was inflated by a small CO_2 cartridge. When fully puffed-up, these bright yellow colored jackets almost resembled the size of the magnificent bosom of the famous cinema star.

7. Statement from Richard W. Lewis, Kenyon crew, 1988.

8. Statement from Clinton A. Bush, Fisher crew, 1988.

9. Radio intercept of German transmission by Allied radio monitoring services, 14 October 1943, USAF Historical Research Agency, Maxwell AFB, Montgomery, Alabama.

10. Ibid.

11. Headquarters, 1st Bombardment Division, *Minutes of Wing and Group Commanders Meeting*, 15 October 1943, p. 2.

12. Budd J. Peaslee, *Heritage of Valor*, (Philadelphia and New York, J.B. Lippincott, 1964) p. 218.

13. Headquarters, 1st Bombardment Division, *Minutes of Wing and Group Commanders Meeting*, 15 October 1943, p. 2.

14. Thomas M. Coffey, *Decision Over Schweinfurt*, (New York, David McKay, 1977), p. 289.

15. Headquarters, 1st Bombardment Division, *Minutes of Wing and Group Commanders Meeting*, 15 October 1943, p. 1.

16. Ibid, p. 2.

17. The entry in the 92d Bomb Group lead navigator's log, 14 October 1943 read, "Large 'S' to fall behind 41st C.W." (which was actually the 1st CBW). The 1st Bombardment Division report to Eighth Bomber Command, dated 20 October 1943, indicated Peaslee made a complete 360 degree turn rather than an "S" turn.

18. Statement from LeRoy V. Pikelis, Kane crew, 1989.

19. Entry in 305th Bomb Group lead navigator's log, 14 October 1943.

20. Ibid.

21. Statement from Clinton A. Bush, Fisher crew, 1988.

22. Statement from Lester J. Levy, Murdock crew, 1988.

23. Statement from Homer L. Hocker, Bullock crew, 1991.

24. Statements from Robert A. Skerry, Skerry crew, 1989 and Gerald B. Eakle, Eakle crew, 1989.

25. Statements from Robert A. Skerry, Skerry crew, 1989, Loren M. Fink, Fisher crew, 1988, and Gerald B. Eakle, Eakle crew, 1989.

26. Entry in 305th Bomb Group lead navigator's log, 14 October 1943.

27. Statement from John C. Lindquist, Skerry crew, 1988.

28. Statement from Loren M. Fink, Fisher crew, 1988.

29. Statement from Frederick B. Farrell, Farrell crew, 1988.

30. Headquarters, Eighth Fighter Command, *Narrative of Operations*, 14 October 1943; and Operation of 14 October 1943.

31. Ibid.

32. Headquarters, 1st Bombardment Division, *Minutes of the Wing and Group Commanders Meeting*, 15 October 1943, p. 1.

33. Entry in 92d Bomb Group lead navigator's log, 14 October 1943.

34. Headquarters, 41st Combat Wing, *Critique of Raid on Schweinfurt, Germany*, 15 October 1943, p. 1.

35. Statement from Gerald B. Eakle, Eakle crew, 1989.

36. Radio intercept of German transmission by Allied radio monitoring services, 14 October 1943, USAF Historical Research Agency, Maxwell AFB, Montgomery, Alabama.

37. Entry in 305th Bomb Group lead navigator's log, 14 October 1943.

38. Thomas M. Coffey, *Decision Over Schweinfurt*, (New York, David McKay, 1977), pp. 288-290, and statement

by George G. Shackley, Colonel, USAF (Ret), 1993.

39. USAF Historical Research Agency, *History of the 1st Combat Wing*, Maxwell AFB, Montgomery, Alabama.

Chapter 6 – Into Holland

1. Headquarters, Eighth Fighter Command, *Narrative of Operations*, 14 October 1943.

2. Ibid.

3. Headquarters, Eighth Bomber Command, (Corrected) *Report of 115th Operation, 14 October 1943, Mission No. 1 – Schweinfurt, Mission No. 2 – Schweinfurt, Mission No. 3 – Diversion*, dated 30 October 1943.

4. Ibid.

5. Ibid.

6. Radio intercept of German transmission by Allied monitoring services, 14 October 1943, USAF Historical Research Agency, Maxwell AFB, Montgomery, Alabama.

7. Statement from Joseph E. Chely, Chely crew, 1989.

8. Ibid.

9. Headquarters 305th Bombardment Group (H), *Abortive Report for Mission of 14 October 1943*, dated 15 October 1943.

10. Intelligence Report from 305th Bombardment Group (H) to Commanding General, 1st Bomb Division, and Commanding Officer, 40th Bomb Wing dated 14 October 1943 provided the coordinates where aircraft #42-30375 turned back.

11. Entries in the 305th Bomb Group lead navigator's log, 14 October 1943.

12. Headquarters, 305th Bombardment Group (H), Office of the Operations Officer, *Desription of Navigator's Problem on the Mission of 14 October 1943*, to the Commanding Officer, 305th Bombardment Group (H), dated 15 October 1943.

13. Headquarters, Eighth Bomber Command, *Enemy Tactics, Schweinfurt – Bomber Command Attack*, 14 October 1943, p. 2.

14. Ibid., p. 3

15. Ibid., p. 3

16. Headquarters, Eighth Fighter Command, *Operation of 14 October 1943*.

17. Headquarters, Eighth Bomber Command, *Enemy Tactics, Schweinfurt – Bomber Command Attack*, 14 October 1943, p. 2.

18. Statement from John A. Cole, Kenyon crew, 1988.

19. Statement from Stanley Alukonis, Lang crew, 1989.

20. Statement from Frederick E. Helmick, Farrell crew, 1988.

21. Headquarters, Eighth Bomber Command, *Enemy Tactics, Schweinfurt – Bomber Command Attack*, 14 October 1943, p. 1.

22. Headquarters, Eighth Bomber Command, *Enemy Tactics Employed Against Our Forces*, 14 October 1943, p. 1.

23. Entry in the 305th Bomb Group lead navigator's log, 14 October 1943.

24. Ibid.

Notes

25. Headquarters, Eighth Fighter Command, *Operation of 14 October 1943*.

26. Ibid.

27. Headquarters, 41st Combat Wing (H), *Critique of Raid on Schweinfurt, Germany*, dated 15 October 1943, p. 2.

28. Statement from Frederick E. Helmick, Farrell crew, 1988.

29. Headquarters, Eighth Bomber Command, *Enemy Tactics Employed Against Our Forces*, 14 October 1943, p. 2.

30. Statement from John C. Tew, Jr., Lang crew, 1988.

31. Headquarters, Eighth Bomber Command, *Enemy Tactics Employed Against Our Forces*, 14 October 1943, p. 2.

32. Statement from Gerald B. Eakle, Eakle crew, 1989.

33. Headquarters, Eighth Bomber Command, *Enemy Tactics Employed Against Our Forces*, 14 October 1943, p. 4.

34. Headquarters, Eighth Bomber Command, *Enemy Tactics, Schweinfurt – Bomber Command Attack*, 14 October 1943, p. 2.

35. Statement from Jayson C. Smart, Farrell crew, 1989.

36. Headquarters, Eighth Bomber Command, *Enemy Tactics, Schweinfurt – Bomber Command Attack,* 14 October 1943, p. 3.

37. Ibid.

Chapter 7 – At the Border

1. Statement from Edwin L. Smith, Murdock crew, in 1988.

2. Ibid.

3. Letter to the author from Ron W.M.A. Putz, World War II Airwar Research Group, The Netherlands, 8 June 1989.

4. Statement from Edwin L. Smith, Murdock crew, 1988.

5. Missing Aircrew Report #917, National Archives, Suitland, Maryland.

6. Ibid.

7. Ibid.

8. Statement from Edwin L. Smith, Murdock crew, 1988.

9. Statement from Lester J. Levy, Murdock crew, 1988.

10. Ibid.

11. Statement from Edwin L. Smith, Murdock crew, 1988.

12. Ibid.

13. German Document AV 390/43, *Report on Losses from 3 September to 20 October 1943*, No. 80, National Archives, Suitland, Maryland.

14. Letter to William B. Menzies, Murdock crew, from Peter H. Luijten, World War II Airwar Research Group, The Netherlands, dated 22 October 1979.

15. Statement from Dennis J. McDarby, McDarby crew, 1989.

16. Robert E. O'Hearn, *In My Book You're All Heroes*, (California, Hall Letter Shop), 1984, 1988, p. 102.

17. Statement from Dennis J. McDarby, McDarby crew, 1989.

18. Statement from Arthur E. Linrud, McDarby crew, 1988.

19. Letter to Dennis J. McDarby, McDarby crew, from Ron W.M.A. Putz, World War II Airwar Research Group, The Netherlands, 27 November 1988.

20. Ibid.

21. Robert E. O'Hearn, *In My Book You're All Heroes*, (California, Hall Letter Shop), 1984, 1988, p. 104.

22. Missing Aircrew Report #1034, National Archives, Suitland, Maryland.

23. Ibid.

24. Robert E. O'Hearn, *In My Book You're All Heroes*, (California, Hall Letter Shop, 1984, 1988, p. 103.

25. Ibid.

26. Statement from Dennis J. McDarby, McDarby crew, 1989.

27. Letter to William B. Menzies, Murdock crew, from Frans Soomers, Eygelshoven, The Netherlands, dated 25 October 1979. This letter was translated by the editors of "The Escape" publication, The Netherlands, and forwarded to Menzies after he had requested information about the Murdock aircraft in their publication. This letter contained some information about the Murdock crew, but mainly concerned the McDarby crew.

28. Statement from Alfredo A. Spadafora, Eakle crew, 1988.

29. Statement from Gerald B. Eakle, Eakle crew, 1989.

30. Statement from Herman E. Molen, Eakle crew, 1988.

31. Statement from Alfredo A. Spadafora, Eakle crew, 1988.

32. Remarks by Lloyd G. Wilson on his Questionnaire for Service Personnel Evading from Enemy Occupied Countries, Headquarters, European Theater of Operations, Gibraltar, 22 January 1944, National Archives, Suitland, Maryland.

33. Ibid.

34. Statement from Gerald B. Eakle, Eakle crew, 1989.

35. Statement from Alfredo A. Spadafora, Eakle crew, 1988.

36. Statement from Gerald B. Eakle, Eakle crew, 1989.

37. Remarks by Lloyd G. Wilson on his Questionnaire for Service Personnel Evading from Enemy Occupied Countries, Headquarters, European Theater of Operations, Gibraltar, 22 January 1944, National Archives, Suitland, Maryland.

38. Statement from Herman E. Molen, Eakle crew, 1988.

39. Remarks by Lloyd G. Wilson on his Questionnaire for Service Personnel Evading from Enemy Occupied Countries, Headquarters, European Theater of Operations, Gibraltar, 22 January 1944, National Archives, Suitland, Maryland.

40. Ibid.

41. Statement from Herman E. Molen, Eakle crew, 1988.

42. Remarks by Lloyd G. Wilson on his Questionnaire for Service Personnel Evading from Enemy Occupied Countries, Headquarters, European Theater of Operations, Gibraltar, 22 January 1944, National Archives, Suitland, Maryland.

43. Ibid.

44. Ibid.

45. Ibid.

46. Ibid.

47. Statement from Robert A. Skerry, Skerry crew, 1989.

48. Ibid.

49. Statement from John C. Lindquist, Skerry crew, 1988.

50. Statement from Robert A. Skerry, Skerry crew, 1989.

51. Missing Aircrew Report #918, National Archives, Suitland, Maryland.

52. Ibid.

53. Statement from John L. Gudiatis, Willis crew, 1990.

54. Ibid.

55. Ibid.

56. Missing Aircrew Report #918, National Archives, Suitland, Maryland.

57. Ibid.

58. Ibid.

59. Letter to the author from Leo Zeuren, World War II Airwar Research Group, Holland, 12 January 1989.

60. German Document AV 390/43, *Report on Losses from 3 September to 20 October 1943*, No. 80, National Archives, Suitland, Maryland.

61. Ibid.

Chapter 8 – Into Germany

1. Entry in the 305th Bomb Group lead navigator's log at 1333 hours, 14 October 1943 as Normand continued to follow the 1st Combat Wing.

2. Ibid.

3. German antiaircraft reports, 14 October 1943, Bundesarchiv-Militarachive, RL 5/147, Federal Republic of Germany provided by Felix Freiherr von Loe, Burg Adendorf, Germany.

4. Entry in the 305th Bomb Group lead navigator's log, 14 October 1943.

5. Headquarters, Eighth Bomber Command, *Enemy Tactics, Schweinfurt – Bomber Command Attack*, p. 4, 14 October 1943.

6. Entry in the 92d Bomb Group lead navigator's log, 14 October 1943.

7. Headquarters, 41st Combat Wing (H), Critique of Raid on Schweinfurt, Germany, dated 15 October 1943.

8. German antiaircrat reports, 14 October 1943, Bundesarchiv-Militarachive RL 5/147, Federal Republic of Germany provided by Felix Freiherr von Loe, Burg Adendorf, Germany.

9. Statement from LeRoy V. Pikelis, Kane crew, 1989.

10. Statement from William C. Heritage, Kincaid crew, 1988.

11. Statement from Joseph K. Kocher, Bullock crew, 1988.

12. Headquarters, Eighth Bomber Command, *Enemy Tactics, Schweinfurt – Bomber Command Attack*, p. 2, 14 October 1943.

13. Statements from John A. Cole, Kenyon crew, 1988, and William C. Frierson, Holt crew, 1989.

14. Statement from Christy Zullo, Dienhart crew, 1989.

15. Statement from William C. Frierson, Holt crew, 1989.

16. Ibid.

17. Statement from Charles E. Blackwell, Holt crew, 1988.

18. Missing Aircrew Report #914, National Archives, Suitland, Maryland.

19. Ibid.

20. Ibid.

21. Statement from William C. Frierson, Holt crew, 1989.

22. Missing Aircrew Report #914, National Archives, Suitland, Maryland.

23. Statement from William C. Frierson, Holt crew, 1989.

24. Statement from Charles E. Blackwell, Holt crew, 1988.

25. Ibid.

26. Ibid.

27. Missing Aircrew Report #914, National Archives, Suitland, Maryland.

28. Ibid.

29. Ibid.

30. Ibid.

31. Statement from William C. Frierson, Holt crew, 1989.

32. Ibid.

33. Missing Aircrew Report #914, National Archives, Suitland, Maryland.

34. Ibid.

35. At this time, an entry in the 305th Bomb Group lead navigator's log showed the lead squadron was at 21,400 feet.

36. Statement from Edward W. Dienhart, Dienhart crew, 1988.

37. Ibid.

38. Statement from Alfredo A. Spadafora, Eakle crew, 1988.

39. Statements from Dienhart crew members, Raymond C. Baus, 1988, and Christy Zullo, 1988.

40. Statement from Bernard Segal, Dienhart crew, 1989.

41. Ibid.

42. Missing Aircrew Report #913, National Archives, Suitland, Maryland.

43. Statements from Dienhart crew members, Raymond C. Baus, 1988, and Christy Zullo, 1988.

44. Statement from Edward W. Dienhart, Dinehart crew, 1988.

45. Statements from Dienhart crew members, Bernard Segal, 1989, and Raymond C. Baus, 1988.

46. Statement from Bernard Segal, Dienhart crew, 1989.

47. Statement from Raymond C. Baus, Dienhart crew, 1988.

48. Statement from Edward W. Dienhart, Dienhart crew, 1988.

49. Ibid.

50. Statement from Brunson W. Bolin, Dienhart crew, 1990.

51. Statement from Bernard Segal, Dienhart crew, 1988.

Chapter 9 – Approaching the Rhine

1. Statement from Frederick B. Farrell, Farrell crew, 1988.

2. Statement from Stanley Alukonis, Lang crew, 1988.

3. Statement from Loren M. Fink, Fisher crew, 1988.

4. Statement from Clinton A. Bush, Fisher crew, in 1988.

5. Missing Aircrew Report #916, National Archives, Suitland, Maryland.

6. Ibid.

7. Ibid.

8. Statement from Clinton A. Bush, Fisher crew, 1989.

9. Ibid.

10. Ibid.

11. Ibid.

12. Missing Aircrew Report #916, National Archives, Suitland, Maryland.

13. Ibid.

14. Statement from Loran M. Fink, Fisher crew, 1988.

15. Missing Aircrew Report #916, National Archives, Suitland, Maryland.

16. Ibid.

17. Ibid.

18. Remarks by Stanley Alukonis on his Questionnaire for Service Personnel Evading from Enemy Occupied Countries, Headquarters European Theater of Operations, Gibraltar, 14 January 1944, Ntional Archives, Suitland, Maryland.

19. Ibid.

20. Ibid.

21. Statement from John C. Tew, Jr., Lang crew, 1988.

22. Statement from Charles J. Groeninger, Lang crew, 1988.

23. Statements from Lang crew members, John C. Tew, Jr., 1988 Charles J. Groeninger, 1988, and Stanley Alukonis, 1988.

24. Remarks by Stanley Alukonis on his Questionnaire for Service Personnel Evading from Enemy Occupied Countries, Headquarters European Theater of Operations, Gibraltar, 14 January 1944, National Archives, Suitland, Maryland.

25. Missing Aircrew Report #915, National Archives, Suitland, Maryland.

26. Statement from Stanley Alukonis, Lang crew, 1988.

27. Statement by John C. Tew, Jr., Lang crew, 1988.

28. Ibid.

29. Missing Aircrew Report #915, National Archives, Suitland, Maryland.

30. Ibid.

31. Statement from Charles J. Groeninger, Lang crew, 1988.

32. Statement from Steve Krawczynski, Lang crew, 1988.

33. Ibid.

34. Certified accurate translation, by Carl W. Davis, of a letter written by Josef Palmen, a German farmer from Puffendorf, Germany, and indorsed by Minister Hermanns, Minister of Loverich and Alsdorf, District of Aachen, to the Robert S. Lang family on 15 February 1947, Casualty and Memorial Affairs Operations Center, Alexandria, Virginia.

35. Statement from John A. Cole, Kenyon crew, 1988.

36. Statements from Kenyon crew members, John A. Cole, 1988, and Ellsworth H. Kenyon, 1989.

37. Statement from Ellsworth H. Kenyon, Kenyon crew, 1989.

38. Bill Stoots, SR CAM/AAFTC, formerly of the Memphis Belle Restoration Association, provided the technical information from B-17G manuals.

39. Statement from Richard W. Lewis, Kenyon crew, 1989.

40. Statement from John A. Cole, Kenyon crew, 1988.

41. Statement from Ellsworth H. Kenyon, Kenyon crew, 1989.

42. Statement from Richard W. Lewis, Kenyon crew, 1989.

43. Missing Aircrew Report #911, National Archives, Suitland, Maryland.

44. Ibid.

45. A position report in the 305th Bomb Group lead naviagator's log at 1353 indicated the 305th, adjacent to Milton's 1st Combat Wing, was now back on the briefied heading for Schweinfurt.

46. Time and distance plotting by the author using the Operational Navigation Chart (ONC E-2) 1:1,000,000, Lambert Conformal Conic Projection, DMACSC, Washington, D.C.

47. Map of the 1st Division's flight path from Cologne south to Bonn taken from Friedhelm Golücke: *Schweinfurt und der strategische Luftkrieg 1943*, (Paderborn 1980), p. 256.

48. Statement from Robert Guarini, Skerry crew, 1988.

49. Statement from Robert A. Skerry, Skerry crew, 1989.

50. Information from the files of the Bundesarchiv- Militararchiv-RL 5/147 provided by Felix Freiherr von Loe, Burg Adendorf, Germany, 1989.

51. Statement from Robert A. Skerry, Skerry crew, 1989.

52. Statement from John C. Lindquist, Skerry crew, 1989.

53. Missing Aircrew Report #919, National Archives, Suitland, Maryland.

54. Ibid.

55. Ibid.

56. Map of the 1st Division's flight path from Cologne south to Bonn taken from Friedhelm Golücke: *Schweinfurt und der strategische Luftkrieg 1943*, (Paderborn 1980), p. 256.

57. Information from the files of the Bundesarchiv-Militararchiv-RL 5/147 provided by Felix Freiherr von Loe, Burg Adendorf, Germany, 1989.

58. Missing Aircrew Report #919, National Archives, Suitland, Maryland.

59. Information from the files of the Bundesarchiv-Militararchiv-RL 5/147 provided by Felix Freiherr von Loe, Burg Adendorf, Germany, 1989.

60. Ibid.

61. Statement from John C. Lindquist, Skerry crew, 1989.

62. Ibid.

63. Information from the files of the Bundesarchiv-Militararchiv-RL 5/147 provided by Felix Freiherr von Loe, Burg Adendorf, Germany, 1989.

64. Statement from Robert A. Skerry, Skerry crew, 1989.

65. Statement from Robert Guarini, Skerry crew, 1989.

66. Ibid.

67. Missing Aircrew Report #919, National Archives, Suitland, Maryland.

68. Statement from Robert Guarini, Skerry crew, 1989.

69. Statement from Wayne D. Rowlett, Skerry crew, 1988.

70. Missing Aircrew Report #919, National Archives, Suitland, Maryland.

71. Ibid.

72. Ibid.

73. Ibid.

74. Statement from Robert A. Skerry, Skerry crew, 1989.

75. Information provided by Felix Freiherr von Loe, Burg Adendorf, Germany, 1989.

Chapter 10 – The Bomb Run

1. A position report in the 305th Bomb Group lead navigator's log at 1353 indicated the 305th, adjacent to Milton's 1st Combat Wing, was now back on the briefed heading for Schweinfurt.

2. Headquarters, Eighth Bomber Command, *Enemy Tactics, Schweinfurt – Bomber Command Attack*, p. 4, 14 October 1943.

3. Position reports between 1351 and 1353 hours, from the lead navigators' logs of the 92d and 305th Bomb Groups, show Peaslee's 40th Combat Wing was now ahead of Milton's 1st CBW.

4. Headquarters, 41st Combat Wing (H), *Critique of Raid on Schweinfurt, Germany*, p. 2, dated 15 October 1943.

5. Radio intercept of German transmission by Allied monitoring services, 14 October 1943, USAF Historical Research Agency, Maxwell AFB, Montgomery, Alabama.

6. During the entire battle, the 1st Combat Wing, spearheading the 1st Division, lost a total of three B-17s to the enemy fighters. Headquarters, Eighth Bomber Command, *Bomber Command Narrative of Operations, 115th Operation, 14 October 1943, Mission No. 1 – Schweinfurt, Mission No. 2 – Schweinfurt, Mission No. 3 – Diversion*; Headquarters, Eighth Bomber Command, to Commanding General, Eighth Air Force, *115th Operation, 14 October 1943*, dated 30 October 1943.

7. Headquarters, Eighth Bomber Command, *Enemy Tactics, Schweinfurt – Bomber Command Attack*, p. 3, 14 October 1943.

8. Position reports between 1415 and 1417 hours, from the lead navigators' logs of the 92d and 305th Bomb Groups, show Peaslee's 40th Combat Wing was still ahead of Milton's 1st CBW.

9. Position report at 1417 hours in the lead navigator's log of the 305th Bomb Group, 14 October 1943.

10. Statement from Roy A. Burton, Farrell crew, 1989.

11. Statement from Jayson C. Smart, Farrell Crew, 1989.

12. Ibid.

13. Ibid.

14. Statement from Frederick B. Farrell, Farrell crew, 1988.

15. Ibid.

16. Headquarters, 305th Bombardment Group (H), Office of the Operations Officer, *Bombing Approach and Dropping used by the 305th Bombardment Group (H), on Day Mission of 14 October 1943*, to the Commanding Officer, 305th Bombardment Group (H), dated 15 October 1943.

17. At this time, the 305th Bomb Group lead navigator's log showed the lead squadron at an indicated altitude of 21,700 feet.

18. Missing Aircrew Report #922, National Archives, Suitland, Maryland.

19. Ibid.

20. Ibid.

21. Ibid.

22. Ibid.

23. Statement from John F. Raines, Kincaid crew, 1989.

24. Statement from Carl J. Brunswick, Bullock crew, 1989.

25. Headquarters, 305th Bombardment Group (H), Office of the Operations Officer, *Bombing Approach and Dropping used by the 305th Bombardment Group (H), on Day Mission of 14 October 1943*, to the Commanding Officer, 305th Bombardment Group (H), dated 15 October 1943.

26. Statement from Jayson C. Smart, Farrell crew, 1989.

27. Headquarters, 1st Bombardment Division, *Minutes of Wing and Group Commanders Meeting*, 15 October 1943, p. 1.

28. Headquarters, Eighth Bomber Command, *Bomber Command Narrative of Operations, 115th Operation, 14 October 1943, Mission No. 1 – Schweinfurt, Mission No. 2 – Schweinfurt, Mission No. 3 – Diversion*; Headquarters, Eighth Bomber Command, to Commanding General, Eighth Air Force, *115th Operation, 14 October 1943*, dated 30 October 1943.

29. Ibid.

30. Ibid.

31. Position reports in the lead navigators' logs of the 92d and 305th Bomb Groups, 14 October 1943.

32. Entries in the 305th Bomb Group lead navigator's log, 14 October 1943.

33. Headquarters, 305th Bombardment Group (H), Office of the Operations Officer, *Bombing Approach and Dropping used by the 305th Bombardment Group (H), on Day Mission of 14 October 1943*, to the Commanding Officer, 305th Bombardment Group (H), dated 15 October 1943.

34. Ibid.

35. Ibid.

36. John R. McCrary and David E. Scherman, *First of the Many*, (London, Robson Books, 1981) p. 218.

37. Ibid.

38. Statements from Joseph Pellegrini, Kane crew, 1990.

39. Ibid.

40. Missing Aircrew Report #921, National Archives, Suitland, Maryland.

41. Statement from John F. Raines, Kincaid crew, 1989.

42. Ibid.

43. Statement from Kenneth R. Fenn, Kincaid crew, 1989.

44. Statement from Alfred C. Chalker, Kincaid crew, 1989.

45. Statement from Louis Bridda, Kincaid crew, 1988.

46. Statement from William C. Heritage, Kincaid crew, 1988.

47. Statement from Kenneth R. Fenn, Kincaid crew, 1989.

48. Statement from Alden C. Kincaid, Kincaid crew, 1988.

49. Statements from Kincaid crew members, Louis Bridda, 1988, and Alfred C. Chalker, 1989.

50. Statement from John F. Raines, Kincaid crew, 1990.

51. Statement from Kenneth R. Fenn, Kincaid crew, 1989.

52. Statement from Alfred C. Chalker, Kincaid crew, 1989.

53. Statement from Alden C. Kincaid, Kincaid crew, 1988.

54. Statement from William C. Heritage, Kincaid crew, 1988.

55. Statements from Kincaid crew members, Kenneth R. Fenn, 1989, and John F. Raines, 1989.

56. Statement from Kenneth R. Fenn, Kincaid crew, 1989.

57. Ibid.

58. Statement from Alden C. Kincaid, Kincaid crew, 1988.

59. Statement from John F. Raines, Kincaid crew, 1989.

60. Missing Aircrew Report #921, National Archives, Suitland, Maryland; and a letter written to Leo Zeuren, Baexem, Holland, from William Donald, Northants, England, dated 27 January 1988.

61. Robert E. O'Hearn, *In My Book You're All Heroes*, (California, Hall Letter Shop, 1984, 1988), pp. 105, 107; and statements from Bullock crew members Brunswick, Kocher, and Jarosynski resolved the following: Raymond P. Bullock, Carl J. Brunswick, Stanley J. Jarosynski, and Alden B. Curtis were all flying their 25th mission. Joseph K. Kocher and Harold E. Coyne were flying their 24th mission.

62. Statement from Stanley J. Jarosynski, Bullock crew, 1988.

63. Statement from Joseph K. Kocher, Bullock crew, 1988.

64. Statement from Carl J. Brunswick, Bullock crew, 1989.

65. Statements from Bullock crew members, Joseph K. Kocher, 1988, Carl J. Brunswick, 1989, and Stanley J. Jarosynski, 1988.

66. Robert E. O'Hearn, *In My Book You're All Heroes*, (California, Hall Letter Shop, 1984, 1988), p. 107, and statements from Joseph K. Kocher, Bullock Crewwww, 1988.

67. Thomas M. Coffey, *Decision Over Schweinfurt*, (New York, David McKay, 1977) p. 305.

68. Statements from Joseph W. Kane and Joseph Pellegrini, Kane crew, in 1990.

69. Headquarters, 305th Bombardment Group (H), Office of the Intelligence Officer, *Teletype Section "A", S-7 report*, to the Commanding General, 1st Bombardment Division and the Commanding Officer, 40th Combat Wing, 14 October 1943.

70. Headquarters, 305th Bombardment Group (H), Office of the Operations Officer, *Narrative Report of Operations of the 305th Bombardment Group (H), SCHWEINFURT, 14 October 1943*, to the Commanding Officer, 305th Bombardment Group (H), dated 15 October 1943.

71. Thomas M. Coffey, *Decision Over Schweinfurt*, (New York, David McKay, 1977) pp. 306-307. Actual name is Joseph Pellegrini.

72. Statements from Joseph W. Kane and Joseph Pellegrini, Kane crew, 1990.

73. Headquarters, 1st Bombardment Division, *Minutes of Wing and Group Commanders Meeting*, 15 October 1943, p. 4.

74. Statement from Joseph W. Kane, Kane crew, 1989.

75. Statements from Roy A. Burton, 1989, Jayson C. Smart, 1989, and Frederick E. Helmick, 1988, of the Farrell crew, and Joseph W. Kane, Kane crew, 1989.

76. Headquarters, 1st Bombardment Division, *Minutes of Wing and Group Commanders Meeting*, 15 October 1943, p. 3.

77. Ibid.

78. Radio intercepts of German transmissions by Allied monitoring services, 14 October 1943, USAF Historical Research Agency, Maxwell AFB, Montgomery, Alabama.

79. Headquarters, 1st Bombardment Division, *Annex II – Bombing, Tactical mission Report, Field Order no. 220, Target – Schweinfurt*, 14 October 1943, and the 92nd Bomb Group's lead navigator's log, 14 October 1943.

80. Headquarters, 305th Bombardment Group (H), Mission Camera Report, 14 October 1943.

81. Headquarters, 1st Bombardment Division, *Minutes of Wing and Group Commanders Meeting*, 15 October 1943, p. 3.

82. Statements from Bullock crew members, Homer L. Hocker, 1991, and Carl J. Brunswick, 1989.

83. Statements from Bullock crew members, Stanley J. Jarosynski, 1988, and Joseph K. Kocher, 1988.

84. Ibid.

85. Statement from Carl J. Brunswick, Bullock crew, 1989.

86. Statement from Homer L. Hocker, Bullock crew, 1991.

Chapter 11 – The Return Home

1. Entry in the 305th Bomb Group lead navigator's log recorded the indicated altitude at 21,000 feet.

2. Headquarters, Eighth Bomber Command, *Operation no. 115, Digest of Field Orders*, 14 October 1943, p. 2; and Headquarters, 305th Bombardment Group (H), *Operational Narrative Check List, Navigator's Flight Plan*, 14 October 1943, Item No. 2.

3. Statement from Frederick E. Helmick, Farrell crew, 1988.

4. Statements from Farrell crew members, Frederick B. Farrell, 1988, Jayson C. Smart, 1989, and Marvin D. Shaull, 1990.

5. Statement from Roy A. Burton, Farrell crew, 1989.

6. Statement from Frederick E. Helmick, Farrell crew, 1988.

7. Ibid.

8. Statement from Roy A. Burton, Farrell crew, 1989.

9. Statement from Frederick E. Helmick, Farrell crew, 1988.

10. Statement from Roy A. Burton, Farrell crew, 1989.

11. Entries in the 305th Bomb Group lead navigator's log.

12. Headquarters, Eighth Bomber Command, *Bomber Command Narrative of Operations, 115th Operation, Mission No. 1 – Schweinfurt, Mission No. 2 – Schweinfurt, Mission No. 3 – Diversion*, 14 October 1943.

13. Ibid.

14. Radio intercepts of German transmissions by Allied monitoring services, 14 October 1943, USAF Historical Research Agency, Maxwell AFB, Montgomery, Alabama.

15. Entry in the 305th Bomb Group lead navigator's log, 14 October 1943.

16. Statement from Frederick B. Farrell, Farrell crew, 1988.

17. Statement from Jayson C. Smart, Farrell crew, 1989.

18. Ibid.

19. Headquarters, Eighth Bomber Command, *Enemy Tactics, Schweinfurt – Bomber Command Attack*, 14 October 1943, p. 4.

20. Statement from Joseph W. Kane, Kane crew, 1989.

21. Headquarters, Eighth Bomber Command, *Enemy Tactics, Schweinfurt – Bomber Command Attack*, 14 October 1943, p. 4.

22. Ibid, pp. 4-5.

23. Entry in the 305th Bomb Group lead navigator's log, 14 October 1943.

24. Headquarters, Eighth Bomber Command, *Operation No. 115, Digest of Field Orders*, 14 October 1943, p. 3.

25. Ibid.

26. Headquarters, 1st Bombardment Division, *Report of Operations, Schweinfurt, Germany, 14 October 1943, Rendezvous with Fighters*, paragraph 2a, to Commanding General, Eighth Bomber Command, dated 20 October 1943.

27. Headquarters, Eighth Bomber Command, *Enemy Tactics, Schweinfurt – Bomber Command Attack*, 14 October 1943, p. 4.

28. Ibid, p. 5.

29. Headquarters, 305th Bombardment Group (H), Office of the Operations Officer, *Narrative Report of Operations of the 305th Bombardment Group (H), Schweinfurt, 14 October 1943*, p. 2, to Commanding Officer of the 305th Bombardment Group (H), dated 15 October 1943.

30. Entry in the 305th Bomb Group lead navigator's log, 14 October 1943.

31. Statement from Bel Guber in 1989. From 1942-1945 Bel was assigned, as a nurse, to the 49th Station Hospital.

32. Entry in the 305th Bomb Group lead navigator's log, 14 October 1943.

33. Eighth Bomber Command Operational Route Forecast for 14 October 1943.

34. The revised weather estimate published later in the day on 14 October 1943 by Headquarters, Eighth Bomber Command.

35. Headquarters, 1st Bombardment Division, *Report of Operations, SCHWEINFURT, Germany, 14 October 1943*, p. 1, to Commanding General, Eighth Bomber Command, dated 20 October 1943.

36. Headquarters, 1st Bombardment Division, *Annex VI, Tactical Mission Report, Field Order No. 220, Schweinfurt*, 14 October 1943,

37. Statement from Roy A. Burton, Farrell crew, 1989.

38. Headquarters, 305th Bombardment Group (H), Office of the Intelligence Officer, *Teletype Section "A", S-9 Section*, 14 October 1943 to Commanding General, 1st Bombardment Division and Commanding Officer, 40th Combat Wing.

39. Ibid.

40. Statement from Joseph W. Kane, Kane crew, 1989.

41. Statement from Frederick E. Helmick, Farrell crew, 1988.

42. Statement from Roger J. Goddard, Farrell crew, 1992.

43. Ibid.

44. Statement from Jayson C. Smart, Farrell crew, 1989.

45. Statement from Joseph Pellegrini, Kane crew, 1990.

46. Ibid.

47. Statement from William G. Bowers, 364th Squadron crew chief, 1991.

Book II – Aftermath

Chapter 12 – The Next Day

1. Examination, by the author in 1989, of mortuary records of deceased 305th Bomb Group personnel from Eighth

Air Force Mission Number 115, at the Casualty and Memorial Affairs Operations Center, Alexandria, Virginia.

2. Ibid.

3. Ibid.

4. Those who made it, automatically became members of the International Caterpillar Club, entitling them to wear the gold Caterpillar Badge. Established in the early 1920's by Leslie Irvin, inventor of the modern parachute, this organization contains only members who have saved their lives by using a parachute. To this day, Irvin Industries Canada, Limited, still continues with the Caterpillar Club tradition.

5. History of the 306th Bombardment Group (H), USAF Historical Research Agency, Maxwell AFB, Montgomery, Alabama.

6. Statement from Joseph W. Kane, Kane crew, 1989.

7. Ibid.

8. Statement from Thomas K. McGehee, Lieutenant General, USAF (Ret), 1990.

9. A periodic, pulp magazine thriller, written by Robert J. Hogan for young men in the 1930s-1940s. It was fantasy, depicting World War I aerial combat. Heroic pilots performed daredevil feats of bravery, as they vanquished evil, supernatural beings, Bowling Green State University, Ohio, 1990.

10. Statements from Joseph W. Kane, Kane crew, in 1989 and 1991.

11. USAF Historical Research Agency, Maxwell AFB, Alabama.

12. Headquarters, 305th Bombardment Group (H), Office of the Intelligence Officer, *Supplement to Teletype, Section "A", covering the mission of 14 October 1943, paragraph S-6*, to Commanding General, 1st Bombardment Division, dated 16 October 1943.

13. USAF Historical Research Agency, Maxwell AFB, Montgomery, Alabama.

14. Headquarters, 305th Bombardment Group (H), *Form "A", Bombardier's Log*, 14 October 1943.

15. Headquarters, 305th Bombardment Group, *Photo & Bomb Plotting Report*, 366th Squadron, 14 October 1943.

16. Ibid.

17. Unless otherwise annotated, the information contained in the author's account of this meeting is based upon Headquarters, 1st Bombardment Division, Office of the Commanding General, *Minutes of Wing and Group Commanders Meeting*, dated 15 October 1943. (Italics by the author).

18. Entry in the 92d Bomb Group lead navigator's log, 14 October 1943.

19. Entry in the 305th Bomb Group lead navigator's log, 14 October 1943.

20. Headquarters, Eighth Bomber Command, *Bomber Command Narrative of Operations, 115th Operation, 14 October 1943, Mission No. 1 – Schweinfurt, Mission No. 2 – Schweinfurt, Mission No. 3 – Diversion*; Headquarters, Eighth Bomber Command, *115th Operation, 14 October 1943*, to Commanding General, Eighth Air Force, dated 30 October 1943.

21. Headquarters, 353d Fighter Group Report to Headquarters, Eighth Fighter Command, 14 October 1943.

22. Wesley Frank Craven and James Lea Cate, *The Army Air Forces in World War II*, Volume II, (The University of Chicago, 1949), p. 705.

23. Headquarters, 305th Bombardment Group (H), Office of the Operations Officer, *Desription of Navigator's Problem on the Mission of 14 October 1943*, to Commanding Officer, 305th Bombardment Group (H), dated 15 October 1943.

24. Ibid.

25. Ibid.

26. Ibid.

27. Headquarters, Eighth Bomber Command, *Bomber Command Narrative of Operations, 115th Operation, 14 October 1943, Mission No. 1 – Schweinfurt, Mission No. 2 – Schweinfurt, Mission No. 3 – Diversion*; Headquarters, Eighth Bomber Command, *115th Operation, 14 October 1943*, to Commanding General, Eighth Air Force, dated 30 October 1943.

28. Headquarters, 305th Bombardment Group (H), Office of the Operations Officer, *Bombing Approach and Dropping used by the 305th Bombardment Group (H) on Day Mission of 14 October 1943*, to Commanding Officer, 305th Bombardment Group (H), dated 15 October 1943.

29. Wesley F. Craven and James L. Cate, *The Army Air Forces in World War II*, Volume II, (The University of Chicago, 1949). Pages 849 and 850 list 166 heavy bombers lost from 2- 14 October 1943 as the official count. This did not include returning crews who bailed out of their aircraft over England, ditched the planes in the channel, crash landed their bombers in England, or B-17 and B-24s that had to be scrapped.

30. Headquarters, Eighth Bomber Command, *Bomber Command Narrative of Operations, 115th Operation, 14 October 1943, Mission No. 1 – Schweinfurt, Mission No. 2 – Schweinfurt, Mission No. 3 – Diversion*.

31. Wesley Frank Craven and James Lea Cate, *Army Air Forces in World War II*, Volume II, (The University of Chicago, 1949), pp. 703-704.

Chapter 13 – The First Twelve Minutes of Battle

1. Statements from William C. Frierson, Holt crew, 1989, and Lester J. Levy, Murdock crew, 1988.

2. German Document AV 390/43, Report on Losses from 3 September to 20 october 1943, No. 80, National Archives, National Archives, Suitland, Maryland.

3. Statement from Lester J. Levy, Murdock crew, 1988.

4. Statement from Edwin L. Smith, Murdock crew, 1988.

5. Ibid.

6. Remarks by Thelma B. Wiggins on his Questionnaire for Service Personnel Evading from Enemy Occupied Countries, Headquarters, European Theater of Operations, Gibraltar, 11 January 1944, National Archives, Suitland, Maryland.

7. Letter from Headquarters, 3060th Quartermaster Graves Registration Company, ETO, to Chief, GR/E Division, ETO, 14 July 1945.

8. Missing Aircrew Report #917, National Archives, Suitland, Maryland.

9. Ibid.

10. Statement from Edwin L. Smith, Murdock crew, 1988.

11. Statement from William C. Menzies, Murdock crew, 1989.

12. Letter from Peter H. Luijten, The Netherlands, to William B. Menzies, Murdock crew, 22 October 1979.

13. Memorandum to the Officer in Charge, Casualty Section, Personnel Actions Branch, Adjutant Generals Office, Washington, D.C., p. 1, 31 January 1949.

14. Letter received by, "The Escape", a Dutch publication, from Frans Soomers, Eygelshoven, The Netherlands, translated by Christine Skeijson on 25 October 1979 and sent to William B. Menzies of the Murdock crew.

15. Statements by Murdock crew members Edwin L. Smith, 1988, Lester J. Levy, 1988, and William C. Menzies, 1989.

16. Memorandum to the Officer in Charge, Casualty Section, Personnel Actions Branch, Adjutant Generals Office, Washington, D.C., p. 1, 31 January 1949.

Notes

17. Letter from Headquarters, 3060th Quartermaster Graves Registration Company, ETO, to Chief, GR/E Division, ETO, 14 July 1945.

18. The examination of Douglas L. Murdock's mortuary records at the Casualty and Memorial Affairs Operations Center, Alexandria, Virginia, by the author in 1989, revealed positive identification of the deceased had been determined by dental records.

19. Letter to Dennis J. McDarby, McDarby crew, from Ron W.M.A. Putz, World War II Airwar Research Group, The Netherlands, 27 November 1988.

20. Letter to the author from Ron W.M.A. Putz, World War II Airwar Research Group, The Netherlands, 1 January 1990.

21. Narrative of Investigation at Geilenkirchen, Germany, on 21-22 December 1948 and 4-5 January 1949 as to the whereabouts of Breeden, Donald P. 2d Lt., 0-745041, who was missing in action on 14 October 1943.

22. Statement from Dennis J. McDarby, McDarby crew, 1989.

23. Letter to the author from Ron W.M.A. Putz, World War II Airwar Research Group, The Netherlands, 1 January 1990.

24. Letter received by, "The Escape", a Dutch publication, from Frans Soomers, Eygelshoven, The Netherlands, translated by Christine Skeijson on 25 October 1979 and sent to William B. Menzies of the Murdock crew.

25. Statement from Dennis J. McDarby, McDarby crew, 1989.

26. Statements from McDarby crew members, Arthur E. Linrud, 1988, and Dennis J. McDarby, 1989.

27. Letter to the author from Ron W.M.A. Putz, World War II Airwar Research Group, The Netherlands, 1 January 1990.

28. Statement from Dennis J. McDarby, McDarby crew, 1989.

29. Robert E. O'Hearn, *In My Book You're All Heroes*, (California, Hall Letter Shop, 1984, 1988), p. 104.

30. Letter to the author from Ron W.M.A. Putz, World War II Airwar Research Group, The Netherlands, 1 January 1990.

31. Missing Aircrew Report #1034, National Archives, Suitland, Maryland.

32. Ibid.

33. Letter to the author from Ron W.M.A. Putz, World War II Airwar Research Group, The Netherlands, 1 January 1990.

34. Missing Aircrew Report #1034, National Archives, Suitland, Maryland.

35. Letter to the author from Ron W.M.A. Putz, World War II Airwar Research Group, The Netherlands, 1 January 1990.

36. Letter received by, "The Escape", a Dutch publication, from Frans Soomers, Eygelshoven, The Netherlands, translated by Christine Skeijson on 25 October 1979 and sent to William B. Menzies of the Murdock crew.

37. Statement from Dennis J. McDarby, McDarby crew, 1989.

38. Statements from McDarby crew members, Arthur E. Linrud, 1988, and Dennis J. McDarby, 1989.

39. Narrative of Investigation at Geilenkirchen, Germany on 20 and 21 December 1948 and 4 and 5 January 1949 by Sergeant Maurice J. Talon, RA-11076828, and Board of Review convened at Headquarters, 7887 Graves Registration Detachment, Operations Division, Liege, Belgium, on 11 April 1951. Casualty and Memorial Affairs Operations Center, Alexandria, Virginia.

40. Statement from Gerald B. Eakle, Eakle crew, 1989.

41. Statement from Alfredo A. Spadafora, Eakle crew, 1988.

42. Ibid.

43. Statement from Herman E. Molan, Eakle crew, 1988.

44. Statement from Gerald B. Eakle, Eakle crew, 1989.

45. Remarks by Lloyd G. Wilson on his Questionnaire for Service Personnel Evading from Enemy Occupied Countries, Headquarters, European Theater of Operations, Gibraltar, 22 January 1944, National Archives, Suitland, Maryland.

46. Ibid. In Wilson's questionnaire, he stated he looked at his watch shortly after landing and the watch indicated 1350 hours.

47. Statement from Gerald B. Eakle, Eakle crew, 1989.

48. Statement from Lester J. Levy, Murdock crew, 1988.

49. Missing Aircrew Report # 912, National Archives, Suitland, Maryland.

50. Statement from Gerald B. Eakle, Eakle crew, 1989.

51. Statements from Eakle crew members, Alfredo A. Spadafora, 1988, and Herman E. Molen, 1988.

52. Remarks by Lloyd G. Wilson on his Questionnaire for Service Personnel Evading from Enemy Occupied Countries, Headquarters, European Theater of Operations, Gibraltar, 22 January 1944, National Archives, Suitland, Maryland.

53. Letter to the author from Leo Zeuren, World War II Airwar Research Group, The Netherlands, 13 March 1990.

54. German Document AV 390/43, *Report on Losses from 3 September to 20 october 1943, No. 80*, National Archives, Suitland, Maryland.

55. Missing Aircrew Report #918, National Archives, Suitland, Maryland.

56. Statement from John L. Gudiatis, Willis crew, 1990.

57. Ibid.

58. Missing Aircrew Report #918, National Archives, Suitland, Maryland.

59. Ibid.

60. Statement from Charles J. Groeninger, Lang crew, 1988.

Chapter 14 – The Next Four Minutes

1. Missing Aircrew Report #914, National Archives, Suitland, Maryland.

2. Statement from Charles E. Blackwell, Holt crew, 1988.

3. Statement from William C. Frierson, Holt crew, 1989.

4. Missing Aircrew Report #914, National Archives, Suitland, Maryland.

5. Statement from Charles E. Blackwell, Holt crew, 1988.

6. Missing Aircrew Report #914, National Archives, Suitland, Maryland.

7. Ibid.

8. Ibid.

9. William L. Shirer, *The Rise and Fall of the Third Reich*, (New York, Simon and Schuster, 1960), p. 120.

10. Statement from William C. Frierson, Holt crew, 1989.

11. Statements from Holt crew members, William C. Frierson, 1989, and Charles E. Blackwell, 1988.

12. Statement from Edward W. Dienhart, Dienhart crew, 1988.

13. Missing Aircrew Report #916, National Archives, Suitland, Maryland.

14. Ibid.

15. Statement from Clinton A. Bush, Fisher crew, 1988.

16. Statement from Loren M. Fink, Fisher crew, 1988.

17. Ibid.

18. Missing Aircrew Report #916, National Archives, Suitland, Maryland.

19. Statement from Loren M. Fink, Fisher crew, 1988.

20. Ibid.

21. Ibid.

22. *International World Atlas*, (Chicago, Rand McNally, 1962), pp. 25-26.

23. Ibid., p. 26.

24. Unless otherwise annotated, the following excerpts were provided by Loren M. Fink, Fisher crew, 1988 and 1992, from his World War II log book which he kept while a prisoner of war.

25. *Maps of Germany, Austria, and Switzerland*, (Chicago, Rand McNally, 1962), p. 16.

26. Ibid.

27. Ibid., p. 17.

28. Sheet "C" of Holland, Belgium, France, and Germany escape map kit, 1:1,000,000, issued to Eighth Bomber Command flyers during World War II. This sheet indicates the only right-of-way east of Loburg towards Wiesenburg dead-ends halfway to Wiesenburg which coincides with the area where the prisoners left the train.

29. Headquarters, Eighth Bomber Command, *Inclosure "F" to Operation No. 115, Digest of Field Orders*, 14 October 1943.

Chapter 15 – The Following Eleven Minutes

1. Letter to the author from Frederick B. Farrell, Farrell crew, 1988.

2. Statements from Lang crew members, Steve Krawczynski, 1989, and Stanley Alukonis, 1989.

3. Statement from Steve Krawczynski, Land crew, 1989.

4. Missing Aircrew Report #915, National Archives, Suitland, Maryland.

5. Remarks by Stanley Alukonis, 14 January 1944, and Steve Krawczynski, 2 February 1944, on their Questionnaires for Service Personnel Evading from Enemy Occupied Countries, Headquarters, European Theater of Operations, Gibraltar, National Archives, Suitland, Maryland.

6. Statement from Charles J. Groeninger, Lang crew, 1988.

7. Missing Aircrew Report # 915, National Archives, Suitland, Maryland.

8. Robert E. O'Hearn, *In My Book You're All Heroes*, (California, Hall Letter Shop, 1984, 1988), p. 127.

9. Statement from John C. Tew, Jr., Lang crew, 1988.

10. Ibid.

11. Statements from Lang crew members, Charles J. Groeninger, 1988, and John C. Tew, Jr., 1988.

12. Statement from John C. Tew, Jr., Lang crew, 1988.

13. Statement from Charles J. Groeninger, 1988.

14. Missing Aircrew Report #915, National Archives, Suitland, Maryland.

15. Ibid.

16. Robert E. O'Hearn, *In My Book You're All Heroes*, (California, Hall Letter Shop, 1984, 1988), p. 127.

17. Certified accurate translation, by Carl W. Davis, of a letter written by Josef Palmen, a German farmer from Puffendorf, Germany, and indorsed by Minister Hermanns, Minister of Loverich and Alsdorf, District of Aachen, to the Robert S. Lang family, on 15 February 1947, Casualty and Memorial Affairs Operations Center, Alexandria, Virginia.

18. Ibid.

19. Ibid.

20. Ibid.

21. Statement from Ellsworth E. Kenyon, Kenyon crew, 1989.

22. Missing Aircrew Report #911, National Archives, Suitland, Maryland.

23. Ibid.

24. Statements from Kenyon crew members, John A. Cole, 1988, and Ellsworth E. Kenyon, 1989.

25. Ibid.

26. Statement from John A. Cole, Kenyon crew, 1988.

27. Statements from Kenyon crew members, John A. Cole, 1988, and Ellsworth E. Kenyon, 1989.

28. Statement from John A. Cole, Kenyon crew, 1988.

29. Statement from Ellsworth E. Kenyon, Kenyon crew, 1989.

30. Statement from Richard W. Lewis, Kenyon crew, 1988.

31. Statement from John A. Cole, Kenyon crew, 1988.

32. Missing Aircrew Report #911, National Archives, Suitland, Maryland.

33. Statements from Kenyon crew members, Richard W. Lewis, 1988, and John A. Cole, 1988.

34. Statement from John A. Cole, 1988.

35. Ibid.

36. Statements from Kenyon crew members, Richard W. Lewis, 1988, John A. Cole, 1988, and Ellsworth E. Kenyon, 1989.

37. Statement from Robert A. Skerry, Skerry crew, 1989.

38. Ibid.

39. Time and distance computations, by the author, placed the 305th Bomb Group's position at 1402 hours approximately 11 miles northeast of Koblenz.

Notes 291

40. Translations of numerous German documents from Bundesarchiv-Militararchive by Felix Freiherr von Loë, Burg Adendorf, Germany, 1989.

41. From the World War II diary kept by John C. Lindquist, Skerry crew, 1988.

42. Ibid.

43. Statement from Robert Guarini, Skerry crew, 1988.

44. Ibid.

45. Statement from Robert A. Skerry, Skerry crew, 1989.

46. Statement from Wayne D. Rowlett, Skerry crew, 1988.

47. Letter to Felix Freiherr von Loë, Burg Adendorf, Germany, from Edward E. DeVaul, Skerry crew, 1 December 1986.

48. Statements from Skerry crew members, Robert Guarini, 1988, John C. Lindquist, 1988, and Robert A. Skerry, 1989.

49. Letter to Felix Freiherr von Loë, Burg Adendorf, Germany, from Edward E. DeVaul, Skerry crew, 1 December 1986; statement from Wayne D. Rowlett, Skerry crew, 1988.

50. Missing Aircrew Report #919, National Archives, Suitland, Maryland.

51. Statement from Robert Guarini, Skerry crew, 1988.

52. Ibid.

Chapter 16 – Into Burg Adendorf

1. Unless otherwise annotated, the information contained in this chapter was provided by Felix Freiherr von Loë, Burg Adendorf, Germany, 1989-1991.

2. Letters to Felix Freiherr von Loë from Skerry crew members, Edwin E. DeVaul, 1 December 1986, Wayne D. Rowlett, 15 December 1986, Robert A. Skerry, 16 December 1986, John C. Lindquist, 18 April 1987, and Robert Guarini, 2 June 1987,

3. Letters to Robert A. Skerry on 31 October 1986 and to the author on 29 February 1992, from Felix Freiherr von Loë.

4. Felix Freiherr von Loë was able to locate and contact the following Skerry crew members: Edwin E. DeVaul, Wayne D. Rowlett, Robert A. Skerry, John C. Lindquist, Robert Guarini, Gus Doumis, and Cecil S. Key.

5. Letter to Robert A. Skerry from Felix Freiherr von Loë, 22 March 1987.

6. Ibid.

Chapter 17 – Then There Were Five

1. Letter to the author from Frederick B. Farrell, Farrell crew, dated 13 August 1988, indicated he joined Normand's lead squadron 10 minutes prior to reaching the IP.

2. At this time, the last 305th B-17 crew to be shot down was the Skerry outfit at 1402 hours.

3. Combat Bombing Flight Record prepared after the mission by the 305th lead bombardier, Lieutenant Joseph Pellegrini, indicates "time of attack" as 1440 hours.

4. Missing Aircrew Report #922, National Archives, Suitland, Maryland.

5. Ibid.

6. Ibid.

7. Statement from John F. Raines, Kincaid crew, 1988, and Carl J. Brunswick, Bullock crew, 1989.

8. Missing Aircrew Report #922, National Archives, Suitland, Maryland.

9. Ibid.

10. Ibid.

11. In Missing Aircrew Report #922, William H. Connelley reported he met Sergeant Whitehead at Dulag Luft in Frankfurt after they were shot down. In 1988, Lester J. Levy, Murdock crew, stated he had seen Herbert S. Whitehead in Stalag 17B during the war. Also in 1988, Charles J. Groeninger, Lang crew, stated he had seen Charles H. Crane in Stalag 17B during the war.

12. Missing Aircrew Report #922, National Archives, Suitland, Maryland.

13. Entry in the 305th Bomb Group lead navigator's log at 1432 hours.

14. Statements from Kincaid crew members, John F. Raines, 1988, William C. Heritage, 1988, and Alden C. Kincaid, 1988.

15. Statement from Alden C. Kincaid, Kincaid crew, 1988.

16. Statement from John F. Raines, Kincaid crew, 1988.

17. Statement from Alden C. Kincaid, Kincaid crew, 1988.

18. Statement from Kenneth R. Fenn, Kincaid crew, 1990.

19. Statement from William C. Heritage, Kincaid crew, 1988.

20. Statement from Louis Bridda, Kincaid crew, 1988.

21. Statements from Kincaid crew members, Alfred C. Chalker, 1989, John F. Raines, 1988, Louis Bridda, 1988, and William C. Heritage, 1988.

22. Statements from William C. Heritage, Kincaid crew, 1988.

23. Statement from Kenneth R. Fenn, Kincaid crew, 1990.

24. Letter to Leo Zeuren, World War II Research Group, Holland, from William Donald, England, dated 27 January 1988.

25. Missing Aircrew Report #921, National Archives, Suitland, Maryland.

26. Ibid.

27. Statements from William C. Heritage, Kincaid crew, 1989.

28. Statements from Kincaid crew members, John F. Raines, 1988, Alfred C. Chalker, 1989, Louis Bridda, 1988, and William C. Heritage, 1988.

29. Unless otherwise annotated, the information contained in the following excerpts of this chapter was provided by Kenneth R. Fenn, Kincaid crew, from a log he kept from 8 April 1945 through 6 June 1945.

30. William L. Schirer, *The Rise and Fall of the Third Reich*, (New York, Simon and Schuster, 1960), p. 121.

31. Ibid., p. 6.

Chapter 18 – The Last Three

1. Statements from Bullock crew members, Carl J. Brunswick, 1989, Stanley J. Jarosynski, 1988, and Joseph K. Kocher, 1988.

2. Statements from Bullock crew members, Homer L. Hocker, 1991, and Joseph K. Kocher, 1988. Carl J. Brunswick of the Bullock crew also made a statement to this effect in his story published in the book, In My Book You're All Heroes, Robert O'Hearn, 1984, 1988, p. 107.

3. Bombardier's Log and Combat Bombing Flight Record prepared by Lieutenant J. Pellegrini, group lead bombardier of the 305th Bomb Group on 14 October 1943.

4. Statement from Carl J. Brunswick, Bullock crew, 1989.

5. Statement from Homer L. Hocker, Bullock crew, 1991.

6. Statement from Joseph K. Kocher, Bullock crew, 1988.

7. Ibid.

8. Statement from Stanley J. Jarosynski, Bullock crew, 1988.

9. Ibid.

10. Unless otherwise annotated, the information contained in the following excerpts of this chapter was provided by statements from Carl J. Brunswick, Bullock crew, 1989. In addition, much of this information came from Brunswick's story which was published in the book, In My Book You're All Heroes, Robert O'Hearn, 1984, 1988, pp. 107-108.

11. Headquarters, 305th Bombardment Group (H), Office of the Intelligence Officer, *Teletype Section "A", S-9 Report*, 14 October 1943 to Commanding General, 1st Bombardment Division and Commanding Officer, 40th Combat Wing.

12. Statement from Frederick E. Helmick, Farrell crew, 1988.

13. General Order 15, Section III, Headquarters, ETOUSA, 10 March 1944, Distinguished Service Cross for 14 October 1943, presentation made by the Commanding Officer, 83d Bomber Training Group, Midland Army Air Field, Midland, Texas, on 27 April 1944.

14. Unless otherwise annotated, the information contained in the following excerpts of this chapter was provided by statements from Bel Kaufman Guber, the widow of Max Guber.

Chapter 19 – The Internees

1. Statement from Edward W. Dienhart, Dienhart crew, 1988.

2. Ibid.

3. Statements from Dienhart crew members, Christy Zullo, 1988, Raymond C. Baus, 1988, and Edward W. Dienhart, 1988.

4. Statement from Edward W. Dienhart, Dienhart crew, 1988.

5. Statement from Bernard Segal, Dienhart crew, 1989.

6. Ibid.

7. Comments by Robert Cinibulk, Dienhart crew, to Edward W. Dienhart, 1989.

8. Statement from Raymond C. Baus, Dienhart crew, 1989.

9. Statement from Bernard Segal, Dienhart crew, 1989.

10. Statements from Edward W. Dienhart, in 1988, indicated he crossed back and forth over a large river several times while making his low level flight to Switzerland.

11. Ibid.

12. Statement from Edward W. Dienhart, Dienhart crew, in 1988.

13. Ibid.

14. Statement from Bernard Segal, Dienhart crew, 1989.

15. Ibid.

16. Missing Aircrew Report #913, National Archives, Suitland, Maryland; and statements from Raymond C. Baus, Dienhart crew, 1988.

17. Statement from Bernard Segal, Dienhart crew, 1989.

18. Statement from Edward W. Dienhart, Dienhart crew, 1988.

19. Statement from Bernard Segal, Dienhart crew, 1989.

20. Comments by Robert Cinibulk to Edward W. Dienhart, 1989.

21. Statement from Edward W. Dienhart, Dienhart crew, 1989.

22. Statement from Bernard Segal, Dienhart crew, 1989.

23. Statement from Edward W. Dienhart, Dienhart crew, 1989.

24. Ibid.

25. Time and distance computations by the author based upon information provided by Dienhart crew members Edward W. Dienhart, Bernard Segal, Christy Zullo, and Raymond C. Baus. In addition, Jean-Pierre Wilhelm of Switzerland also provided information to confirm times and places of the route flown.

26. Statement from Jean-Pierre Wilhelm of Switzerland in 1989.

27. Statement from Christy Zullo, Dienhart crew, 1988.

28. Statement from Edward W. Dienhart, Dienhart crew, 1989.

29. Ibid.

30. Headquarters, Eighth Bomber Command, *Enemy Tactics, Schweinfurt – Bomber Command Attack*, p. 4, 14 October 1943.

31. Statement from Edward W. Dienhart, Dienhart crew, 1988.

32. Statement from Jean-Pierre Wilhelm, of Switzerland, 1990.

33. Statement from Edward W. Dienhart, Dienhart crew, 1988.

34. Ibid.

35. Statements from Dienhart crew members, Bernard Segal, 1989, and Raymond C. Baus, 1988.

36. Statement from Jean-Pierre Wilhelm of Switzerland, 1990.

37. Ibid.

38. Statement from Edward W. Dienhart, Dienhart crew, 1988.

39. Statement from Raymond C. Baus, Dienhart crew, 1988.

40. Ibid.

41. Statement from Brunson W. Bolin, Dienhart crew, 1990.

42. Unless otherwise annotated, the information contained in the following excerpts of this chapter concerning the funeral of Lieutenant Donald T. Rowley was provided by Raymond C. Baus, Dienhart crew, from selected pages of a diary he kept from 14 October 1943 to 12 November 1944.

43. Statement from Jean-Pierre Wilhelm, Switzerland, 1990.

44. Statement from Edward W. Dienhart, Dienhart crew, 1990.

45. Unless otherwise annotated, the information contained in the following excerpts concerning Adelboden was provided by Edward W. Dienhart, Dienhart crew, 1989.

46. Unless otherwise annotated, the information contained in the following excerpts concerning the "paroles" was provided by Dienhart crew members, Bernard Segal, 1989, Raymond C. Baus, 1988, and Christy Zullo, 1988.

47. Statement from Edward W. Dienhart, Dienhart crew, 1989.

48. Statements from Christy Zullow, Diehart Crew, 1989.

49. Statement from Jean-Pierre Wilhelm, Switzerland, 1990.

Chapter 20 – The Escapees

1. Unless otherwise annotated, the information contained in the following excerpts of Herman Molen's escape was provided by him in an interview conducted by the author in 1988.

2. Unless otherwise annotated, the information contained in the following excerpts of John L. Gudiatis' escape was provided by him in an interview conducted by the author in 1990.

3. Unless otherwise annotated, the information contained in the following excerpts of Charles J. Groeninger's escape plans was provided by him in an interview conducted by the author in 1989.

Chapter 21 – The Evaders

1. Unless otherwise annotated, the following excerpts of the escape and evasion by Thelma B. Wiggens, Jr., were taken from his Questionnaire for Service Personnel Evading from Enemy Occupied Countries, Headquarters, European Theater of Operations, Gibraltar, 17 January 1944, National Archives, Suitland, Maryland.

2. Unless otherwise annotated, the following excerpts of the escape and evasion by Lloyd G. Wilson were taken from his Questionnaire for Service Personnel Evading from Enemy Occupied Countries, Headquarters, European Theater of Operations, Gibraltar, 22 January 1944, National Archives, Suitland, Maryland.

3. Unless otherwise annotated, the following excerpts of the escape and evasion by Howard J. Keenan were provided by Steve Krawczynski.

4. Missing Aircrew Report #915, National Archives, Suitland, Maryland.

5. Ibid.

6. Unless otherwise annotated, the information contained in the following excerpts of Steve Krawczynski's escape and evasion was provided by him in an interview conducted in 1989 and from his Escape and Evasion Questionnaire, located in the National Archives, Suitland, Maryland, which he filled out in Gibraltar on 4 February 1944.

7. Unless otherwise annotated, the information contained in the following excerpts of Stanley Alukonis' escape and evasion was provided by him in an interview conducted in 1989 and from his Escape and Evasion Questionnaire, located in the National Archives, Suitland, Maryland, which he filled out in Gibraltar on 14 January 1944.

Chapter 22 – The Turning Point

1. Headquarters, Eighth Bomber Command, *Bomber Command Narrative of Operations, 115th Operation, 14 October 1943, Mission No. 1 – Schweinfurt, Mission No. 2 – Schweinfurt, Mission No. 3 – Diversion.*

2. Wesley Frank Craven and James Lea Cate, *The Army Air Forces in World War II*, Volume II, (University of Chicago, 1949), pp. 703-704.

3. Headquarters, Eighth Bomber Command, *Bomber Command Narrative of Operations, 115th Operation, 14*

October 1943, Mission No. 1 – Schweinfurt, Mission No. 2 – Schweinfurt, Mission No. 3 – Diversion, dated 30 October 1943.

4. Headquarters, 1st Bombardment Division, *Report of Operations, SCHWEINFURT, Germany, 14 October 1943*, p. 1, to Commanding General, Eighth Bomber Command, dated 20 October 1943.

5. Headquarters, 1st Bombardment Division, *Annex VI, Tactical Mission Report, Field Order No. 220, Schweinfurt*, 14 October 1943.

6. Headquarters, Eighth Fighter Command, Operation of 14 October 1943.

7. Headquarters, Eighth Bomber Command, *Bomber Command Narrative of Operations, 115th Operation, 14 October 1943, Mission No. 1 – Schweinfurt, Mission No. 2 – Schweinfurt, Mission No. 3 – Diversion*.

8. Headquarters, Eighth Bomber Command, *Bomber Command Narrative of Operations, Day Operations – 17 August 1943, Mission No. 84*.

9. Ronald H. Bailey and the Editors of Time-Life Books, *The Air War in Europe*, (Virginia, Time-Life Books, 1979), p. 132.

10. Ibid, p. 135.

11. Wesley Frank Craven and James Lea Cate, *The Army Air Forces in World War II*, Volume II, (University of Chicago, 1949), pp. 719, 849-850.

12. Ibid, p. 705.

13. Ibid.

14. Ibid, pp. 705-706.

Chapter 23 – Epilogue

1. Statement by Thomas K. McGehee, LTG USAF (Ret), in 1990.

2. Headquarters, 1st Bombardment Division, *Report of Operations, SCHWEINFURT, Germany, 14 October 1943*, p. 2, to Commanding General, Eighth Bomber Command, dated 20 October 1943.

3. National Personnel Records Center, Military Personnel Records, St. Louis, Missouri.

4. Ibid.

5. Statements from Charles Sackerson, 305th Bomb Group, in 1989, indicated he was present at the 1st Bombardment Division Headquarters shortly after the 14 October 1943 raid to Schweinfurt. While at headquarters, he witnessed the award of the Silver Star to Normand. The National Personnel Records Center in St. Louis, Missouri confirmed Charles G.Y. Normand was awarded a Silver Star decoration. However, the Center was unable to verify the date of that award as item 42, Awards and Decorations, of Normand's AGO Form 66-2, AAF Officers' Qualification Record had been damaged by fire.

6. War Diary for August 1944, sheet 5, 365th Bomb Squadron, 305th Bomb Group (H) indicates on 24 August 1944 the 365th Squadron lost three B-17s. One of the 30 crew members reported as "missing in action" was Lieutenant Colonel Charles G.Y. Normand. Gerald B. Eakle, pilot of the Eakle crew who was shot down on 14 October 1943, stated in 1989 Lieutenant Colonel Normand was assigned a billet across the hall from him while both were incarcerated in Stalag Luft 3 during the war.

Bibliography

Bailey, Ronald H., *The Air War in Europe*, Time-Life Books, Virginia, 1979.

Caidin, Martin, *Black Thursday*, E.P. Dutton, New York, 1960.

Coffey, Thomas M., *Decision Over Schweinfurt*, David McKay, New York, 1977.

Craven, Wesley Frank, and Cate, James Lea, *The Army Air Forces in World War II*, Volumes I and II, The University of Chicago Press, Chicago, 1949.

Golucke, Friedhelm, *Schweinfurt und der strategische Luftkrieg, 1943*, Paderborn, 1980.

Jablonski, Edward, *Flying Fortress*, Doubleday, New York, 1965.

McCrary, John R., and Scherman, David E., *First of the Many*, Robson Books, London, 1981.

O'Hearn, Robert E., *In My Book You're All Heroes*, Hall Letter Shop, Bakersfield, CA, 1984, (reprint) 1988.

Peaslee, Budd J., Colonel, USAF (Ret.), *Heritage of Valor*, J.B. Lippincott, Philadelphia and New York, 1964.

Shirer, William L., *The Rise and Fall of the Third Reich*, Simon and Schuster, New York, 1960.

Siefring, Thomas A., *U.S. Air Force in World War II*, Chartwell Books, New Jersey, 1977.

305th Bomb Group Combat Crew Handbook, October 1943, rev. 2 February 1945.

International World Atlas, Rand McNally, Chicago, 1962.

Index

Aachen, Germany, 67, 72, 85, 88, 97, 98, 161, 163, 165, 230; P-47 turnaround point, 17
Abbeville, France, 127
Adamson, Silas W., TSgt., top turret, Maxwell crew, 108, 176, 259, 265
Adcox, James G., 2nd Lt., bombardier, Lang crew, 162-163
Adelboden, Switzerland, 206, 208-210
Adendorf, Germany, 167, 169-170, 173
Aesch, Switzerland, 200, 202
Ahlgren, Russell R., TSgt., radio operator, Kenyon crew, 98, 164, 166, 258, 263
Aldis Lamp, 26. 43-44
Almquist, Reuben B., SSgt., top turret, Lang crew, 95-96, 162-163, 224, 258
Altheim, Austria, 182
Alukonis, Stanley, 2nd Lt., copilot, Lang crew, 68, 95-96, 161, 163, 229-245, 258
Amazon River, 245
Amiens, France, 125, 127
Amsterdam, Holland, 151
Anderson, Frederick L., Brig. Gen., CG, Eighth Bomber Command, 135, 137-140, comments at 1st Division critique, 142-143; Schweinfurt messages, 38, 143
Anderson, Orville A., Brig. Gen., D/Comdr. for opns., 135, 137, 139, 142
Anklam, Germany, 143
Antwerp, Belgium, 67
Antwerpen area, Belgium, 67
Army Graves Registration Service, 147
Arnold, Henry H., CG, Army Air Forces, 15, 249
Aurolzmunster, Austria, 182
Automatic pilot (AFCE), 109-110, 142, 166

Baexem, Holland, 80
Ball, Edward O., 2nd Lt., navigator, Holt crew, 86-87, 154, 257, 263
Balsley, J.D., TSgt., top turret, Chely crew, 66
Baltic Sea, 217, 225
Barrett, Bryce, 2nd Lt., bombardier, Holt crew, 86-87, 154, 257, 263
Barth prison camp, 225
Basel, Switzerland, 202, 204
Battle of Britain, 18
Baus, Raymond C., SSgt., ball turret, Dienhart crew, 43, 89-90, 197, 203, 207-209, 257
Bavaria, Germany, 200
Bay of Biscay, 242-244
Beachy Head, England, 159
Bedford, England, 193
Belem, Brazil, 245
Bennett, Harvey, TSgt., radio operator, Fisher crew, 93, 94, 156, 257
Berlin, Germany, 117
Bern, Switzerland, 206, 210
Bialogard, Poland, 156
Bitterfeld, Germany, 159
Bitton, Clinton L., TSgt., top turret, Fisher crew, 94, 156, 257
"Black Thursday", 49. See Caiden, Martin
Blackwell, Charles E., SSgt., top turret, Holt crew, 86-87, 153-154, 257
Blalock, George H., Jr., TSgt., top turret, Dienhart crew, 90, 195, 204, 211, 257
Blasig, Phillip A., 2nd Lt., bombardier, Kincaid crew, 112-113, 177-178, 259, 265
Boeing Aircraft Company, 21
Boggs, Walter H., 2nd Lt., copilot, Eakle crew, 78-79, 150, 256, 261
Bolin, Brunson W., 2nd Lt., copilot, Dienhart crew, 90, 195, 204, 211, 257
Bombers:
  American: B-17F&G, "Flying Fortress", armament, 21; assigned to 1st and 3rd Divisions, 21; ball turret, 25; bombardier position, 23; bomb bay, 24; clock system, 21; cockpit area, 22; engines, 22; navigator position, 23; oxygen system, 26-27; radio compartment, 24; speeds & loads, 22; tail section, 25-26; top turret, 23; waist section, 25; B-18A, submarine patrol, 34; B-24, "Liberator", assigned to 2nd Division, 21;
  British: 4-engine Lancaster, 239; twin-engine Wellington, 42
Bomber stream, 31, 35, 54, 61
Bombing, American daylight, 15-18
Bombing, British night, 16, 165. See Royal Air Force
Bombing, times and altitudes, 31
Bonn, Germany, 83, 85, 98, 100, 103, 109, 117, 137, 166-167, 169-170
Bonones, Luther, SSgt., left waist, Farrell crew, 124, 254
Booth, Carl H., Jr., 2nd Lt., navigator, Fisher crew, 94, 155, 257, 263-264
Bordeaux, France, 241
Bowman, Doris O., TSgt., radio operator, Holt crew, 87-88, 154, 257, 263
Braunau, Austria, 185
Breeden, Donald P., 2nd Lt., copilot, McDarby crew, 76, 148-149, 256, 261, 264
Bremen, Germany, 16-17, 117, 143
Bridda, Louis, Sgt., left waist, Kincaid crew, 112-113, 177-178, 259
Brinkmann, Helmut F., Feldwebel (Sgt.), 76
British Air Sea Rescue, 56-58
British "escapees", 207-208
British Guiana, 245
British Intelligence officer, 215
British reconnaissance unit, 218
Brunswick, Carl J., TSgt., top turret, Bullock crew, 30, 53, 108, 114, 119 189-190, 260
Brussels, Belgium, 125, 233-234, 236
Bullock crew, 114, 187, 260
Bullock, Raymond P., 2nd Lt., pilot, Bullock crew, 30, 42, 44, 46, 53, 61, 85, 105, 108, 114-116, 119, 169, 173, 187, 189, 260
Buncher #40, 35, 46
Burton, Roy A., 2nd Lt., copilot, Farrell crew, 43, 106, 121-122, 254
Bush, Clinton A., 2nd Lt., copilot, Fisher crew, 57-58, 61, 93-94, 155-156, 257

Caiden, Martin, author, 49-50. See "Black Thursday"
Campbell, Ian, 239-244

Index

Camp 11A, 159. See Tent City
Camp International, 213-214
Caribbean Sea, 245
Celle, Germany, 218
Chalker, Alfred C., Sgt., right waist, Kincaid crew, 113, 177-178, 259
Chalking, 41
Chelveston, England, 47, 49, 55, 66, 98, 134, 145, 165, 175, 191, 210, 224, 251; home of 305th, 19; return of group, 127-128
Chely, Joseph E., 1st Lt., pilot, Chely crew, 44-47, 66-67, 106, 175
Chevrolet, European model, 227
Chorley, England, 191
Cinibulk, Robert, SSgt., right waist, Dienhart crew, 89-90, 195-197, 203, 211, 257
Citron, Alan B., SSgt., radio operator, Willis crew, 18, 79, 152, 256
Clacton, England, 48
Clinton Street, New Britain, CT., 228
Coffey, Thomas M., author, 115-116. See, "Decision over Schweinfurt"
Cole, John A., 2nd Lt., navigator, Kenyon crew, 68, 97-98, 164-166, 258
Collins, Joseph F., Jr., 2nd Lt., bombardier, Kenyon crew, 42, 98, 164, 166, 258
Cologne, Germany, 83, 85, 89, 98-100, 105, 109, 137, 196
Combat box formation, 31-32
Combat Tour, 18, 29
Combined Chiefs of Staff (CCS), 15
Committee of Operation Analysts (COA), 15
Congressional Medal of Honor, 134
Conley, Craig T., SSgt., ball turret, Maxwell crew, 176, 259, 265
Connelley, William H., SSgt., tail gunner, Maxwell crew, 108, 175, 176, 259
Courland Lagoon (Kurisches Haff), Lithuania, 156
Courtrai, Belgium, 125
Coyne, Harold E., SSgt., right waist, Bullock crew, 114, 188, 260
Cramer, Christian W., 2nd Lt., navigator, Willis crew, 79, 152, 256
Cranc, Charles H., Sgt., right waist, Maxwell crew, 108, 175-176, 259
Crawford, Hosea F., SSgt., radio operator, McDarby crew, 76, 148-149, 256
Curtis, Alden B., SSgt., tail gunner, Bullock crew, 30, 188, 260

Dakar, Senegal, 245
Danube River, Austria, 180, 181, 190
Daventry, England, 49-59, 140-141
Davis, B.T., SSgt., ball turret, Kane crew, 129, 254
Davis, Thomas H., 2nd Lt., copilot, Kenyon crew 98, 164, 166, 258
D-Day, 6 June 1944, 207
"Decision over Schweinfurt", 115-116. See Coffey, Thomas M.
Deelen sector, south central Holland, 68
DeVaul, Edwin E., TSgt., tail gunner, Skerry crew, 99, 100, 103, 168, 259
Diddington, England, 135, 191-192
Dienes, Tony E., SSgt., left waist, Murdock crew, 61, 74, 145-147, 255, 261

Dienhart crew, 88, 155, 195, 257
Dienhart, Edward W., 2nd Lt., pilot, Dienhart crew, 43, 77, 85, 88-91, 97, 104, 201, 202-204, 206-210
Distinguished Service Cross, 134, 191
Divisions (numbered):

1st Bombardment Division (1st Division), 19, 31, 48-50, 52-56, 58-59, 61-63, 65-70, 72, 83, 98, 105, 107, 109, 111, 117-118-, 123-125, 127, 136, 138-143, 154, 161, 170, 252; assigned aircraft, 21; battle damage, 248; bombers launched, losses, 247-248; bomb loading, 29; escort by 353d Fighter Group, 33; initial point, 36, 107-109; meeting at division headquarters, 15 October 1943, 134, 135-143; Report of Operations, 20 October 1943, 49-50
2nd Bombardment Division (2nd Division), 21, 31, 48, 53, 56, 60, 247
3rd Bombardment Division (3rd Division), 21, 31, 48, 53, 56, 60, 67, 70, 124, 136, 143, 247-248
13th Armored Division, US Third Army, 185

Dixon, Willard E., 2nd Lt., bombardier, Willis crew, 79, 152, 256
Domburg, Holland, 65
Doumis, Gus, SSgt., left waist, Skerry crew, 103, 167-168, 259
Duren, Germany, 97-98, 165
Dusseldorf, Germany, 154, 156
Dutch Guiana, 245
Dutch quislings, 225

Eaker, Ira C., Maj. Gen., CG, Eighth Air Force, 143
Eakle crew, 76, 149, 256
Eakle, Gerald B., 2nd Lt., pilot, Eakle crew, 30, 42, 63, 77-79, 89, 98, 149-151, 213, 221, 256
Edwards, Jack J., 1st Lt., navigator, Kane crew, 48, 50, 53-54, 63, 70, 141, 142, 254
Eighth Air Force, 16-19, 133; directive to abort, 59; loses air superiority, 249; major combat units, 15; mission #115, 19, see Second Schweinfurt; strategic plan for Schweinfurt, 31
Eighth Bomber Command, 17, 19, 29, 35, 139, 142-143, 251; bomber losses 1st Schweinfurt raid, 249; bomber losses 2nd Schweinfurt raid, 247; cancels deep raids, 250; claimed German fighter losses, 249; composition of 16, 21; crew losses 15 September-14 October 1943, 249; estimate of damage, 247
Eighth Fighter Command, 139
Eisden, Belgium, 79, 150, 222
Emperor, John J., 2nd Lt., copilot, Willis crew, 80, 151, 256, 261
Englehardt, Arthur, SSgt., ball turret, Kenyon crew, 98, 164, 166, 258
Enschede-Leeuwarden area, northern Holland, 68
Escape Committee, 213
Escape purse (kit), 189, 222
Escape and Evasion routes, 226
Eygelshoven, Holland, 76, 147-148

Farnborough, England, 224
Farrell crew, 106, 121, 190, 254
Farrell, Frederick B., 1st Lt., pilot, Farrell crew, 30, 35, 43, 45, 47, 53, 68, 70, 72, 93, 98, 103, 107-109, 114, 116, 118-119, 122, 124, 128-129, 134, 135, 161, 175, 187, 191, 254

Fenn, Kenneth R., SSgt., ball turret, Kincaid crew, 112-113, 177-183, 185, 259
Fighters:
American, P-47, "Thunderbolt", as escort, 33; external fuel tanks, 16; P-51, "Mustang" answer to long-range, 250
British Spitfire, 33, 43, 56-57, 125; Polish pilots, 43
German, single-engine: Focke Wulf, Fw 190, Messerschmitt, Bf(Me) 109; twin-engine: Junkers, Ju 88, Messerschmitt, Bf(Me) 110, Messerschmitt, Me 210, 36-37; tactics, 37
Fink, Loren M., SSgt., left waist, Fisher crew, 62, 93-94, 155-157, 258
Fink, William, 155
Finkenrath, Germany, 147
Fisher crew, 93, 155, 257
Fisher, Verl D., F/O, pilot, Fisher crew, 57, 61-62, 93, 95, 98, 150, 156, 257
Ford, V-8, European model, 232
49th Station Hospital, England, 129, 135, 191-193; primary Eighth Air Force facility, 127
Frankfurt, Germany, 105, 106, 107
Frankfurt Interrogation Center, Germany, 166, 190
Free French, 207-208
French Guiana, 245
Frierson, William C., Sgt., ball turret, Holt crew, 29, 53, 86-88, 153-154, 257
Frisian Islands, Holland, 56, 247
Frutigen, Switzerland, 208

G-8 and His Battle Aces, 134
Geilenkirchen, Germany, 156
Geneva Convention, 166
Geneva, Switzerland, 206, 209
German Air Force, 15, 19, 249. See Luftwaffe
German radio traffic, 48, 58, 63, 66, 67, 68, 71, 72, 83, 85, 105, 117, 124, 125, 127
Gestapo, 225-226
Gdynia, Poland, 143
Gibraltar, 224, 227, 229, 244
Goddard, Roger J., TSgt., radio operator, Farrell crew, 129, 254
Gottshall, Walter L., SSgt., tail gunner, Kenyon crew, 98, 166, 258
Grand Atlas Mountains, Morocco, 245
Grand Hotel, Switzerland, 208
Graz, Austria, 214-215
Great Ash Airfield, England, 55
Green, Charles M., SSgt., left waist, Kenyon crew, 97-98, 166, 258
Groeninger, Charles J., SSgt., left waist, Lang crew, 18, 30, 95-96, 162-163, 218-219, 258
Gross Tychow (Tychery), Poland, 156-157
Gross, William M., Col., CO, 1st Combat Wing, 136-139
Groups, American (numbered):
91st Bomb Group, 48, 59, 63, 65, 73, 109, 111, 115-118, 127, 135, 141, 175, 176; bomber losses first Schweinfurt raid, 36; a group of 1st Combat Wing, 31; wing leader on second Schweinfurt, 31
92nd Bomb Group, 36, 47-50, 52-53, 59-60, 83, 109, 117, 127, 138, 248, 252; a group of 40th Combat Wing, 31; wing and division leader on second Schweinfurt, 31
100th Bomb Group, 143

303rd Bomb Group, 59, 118; a group of 41st Combat Wing, 31;
305th Bomb Group, 18-19, 29, 47-50, 52-53, 54-56, 58-61, 63-64, 65-68, 72, 73, 83, 85, 93, 98, 106, 107, 109-111, 114-118, 121, 124, 128-130, 138-141, 142-143, 153, 166, 175, 176, 187, 190-192, 244, 248, 251-252; air battle begins, 145; aircraft for mission, 33; assigned B-17s, 21; bombardier's report, 142; bomber losses first Schweinfurt raid, 36; bomb loading, 29; briefing of mission, 35-38; cameras for bomb run, 118; composition for mission, 33; crew integrity, 30, enemy fighter tactics, 70-71; group lead rotation, 34; a group of 40th Combat Wing, 31; intelligence report of bomb run, 116; low group of wing, 36; navigator's report, 140-141; nickname, 30, operations officer's report of bomb run, 116; ordnance officer's report, 135; pictures of bomb run, 135
306th Bomb Group, 47-50, 59, 65, 83, 109, 134, 138, 143, 248, 252; a group of 40th Combat Wing, 31; high group of wing, 36
351st Bomb Group, 59, 63-64, 65, 109, 137, 141; a group of 1st Combat Wing, 31; high group of wing, 48
352nd Fighter Group, 247
353rd Fighter Group, 33, 63, 66, 248-249; composition of, 65; limit of escort, 68
379th Bomb Group, 59, 70; a group of the 41st Combat Wing, 31; wing leader on second Schweinfurt, 31
381st Bomb Group, 47-48, 59, 63-64, 65, 109, 117, 136-137, 141; bomber losses first Schweinfurt raid, 36; a group of 1st Combat Wing, 31; low group of wing, 53; medical log entry on morale, 17-18;
384th Bomb Group; a group of 41st Combat Wing, 31; low group in wing, 59
Groups, German (numbered):
Jagdgeschwader 1, 76
Jagdgeschwader 2, 99
Jagdgeschwader 26, 74, 127, 167
Guarini, Robert, 2nd Lt., navigator, Skerry crew, 100, 167-168, 259
Guber, Max, 2nd Lt., navigator, Farrell crew, 43, 121-123, 128-129, 135, 191-193, 254
Gudiatis, John L., SSgt., right waist, Willis crew, 79-80, 151-152, 216, 217, 256
Guild Hall, London, England, 193

Haanrade, Holland, 76
Halle, Germany, 159
Hannover, Germany, 157, 218
Hanson, Owen R., TSgt., radio operator, Kane crew, 129, 254
Hawaii, 42, 111
Heerlen, Holland, 225
Helmick, Frederick E., 1st Lt., bombardier, Farrell crew, 30, 43, 53, 68, 70, 121-123, 128-129, 135, 191, 254
Henlin, Leonard R., Sgt., left waist, McDarby crew, 76, 148, 256, 261
"Heritage of Valor", 59. See Colonel Budd J. Peaslee
Heritage, William C., Sgt., tail gunner, Kincaid crew, 30, 112-113, 177-178, 259
Hermanns, Herr, 163
Heydekrug, Lithuania, 156
Higdon, James F, SSgt., ball turret, Farrell crew, 124, 254
Hissom, Donald L., SSgt., togglier, Fisher crew, 94,

Index

195, 257, 263
Hitler, Adolf, 185
Hitler Youth, 154
Hocker, Homer L., 2nd Lt., copilot, Bullock crew, 61, 119, 187-188, 260
Hoisdorf, Austria, 182
Holland-Ruhr area, 58
Holt crew, 86, 153, 257
Holt, Robert W., 2nd Lt., pilot, Holt crew, 29, 52-53, 85, 86, 89, 91, 97, 154, 161, 164, 257, 263
Holy Virgin, 173-174
Horn, Holland, 80-81, 151
Hornli cemetery, Switzerland, 204-205

Immendorf, Germany, 88, 154
Insdorf, Harold, SSgt., ball turret, Fisher crew, 93, 155, 258, 263
Italy, 207

Jackson, Harold E., SSgt., radio operator, Bullock crew, 189, 260
Jamaica, 245
Japanese, 42, 112, 193
Jarosynski, Stanley J., SSgt, left waist, Bullock crew, 30, 114, 119, 188-189, 260
Jausen, Mr., Netherlands, 146
Jessen, Germany, 159
Johnson, Carl A., 2nd Lt., bombardier, Dienhart crew, 89, 196-197, 202-203, 206, 210-211, 257
Johnson, Jack G., SSgt., right waist, Skerry crew, 103, 167, 173, 259, 263
Jones, Charles B., 2nd Lt., navigator, Eakle crew, 77-78, 150-151, 256

Kaeli, Walter A., SSgt., togglier, Bullock crew, 188, 260
Kallham, Austria, 182
Kane crew, 116, 190, 254
Kane, Joseph W., 2nd Lt., pilot, Kane crew, 33, 42, 45-47, 52, 54, 55, 60-61, 115-117, 119, 124, 128-129, 134, 187, 191, 254
Karns, Floyd J., TSgt., top turret, Willis crew, 80, 152, 256
Kassel, Germany, 16
Kaufman, Bel, 2nd Lt., 49th Station Hospital, England, 191-193
Keenan, Howard J., SSgt., right waist, Lang crew, 96, 161, 163, 224-226, 230, 258
Kenyon crew, 97, 164, 258
Kenyon, Ellsworth H., 2nd Lt., pilot, Kenyon crew, 18, 35, 42, 45, 47, 53, 56, 63, 68, 70, 85, 89, 91, 97-99, 161, 164-166, 195, 258
Key, Cecil S., Sgt., togglier, Skerry crew, 100, 168, 259
Kiel, Germany, 42, 155
Kiggens, Russell J., TSgt., top turret, Murdock crew, 75, 146-147, 255, 261
Kincaid crew, 111, 176, 259
Kincaid, Alden C., 2nd Lt., pilot, Kincaid crew, 30, 42, 45, 85, 105, 107, 108, 109, 111-114, 176-178, 187, 259
Klamm, Austria, 180
Klister, Urban H., 2nd Lt., navigator, Maxwell crew, 108, 176, 259, 265
Knapp, Lloyd F., Jr., SSgt., right waist, Eakle crew, 77-78, 149-150, 256
Koblenz, Germany, 67, 168

Kocher, Joseph K., SSgt., ball turret, Bullock crew, 30, 42, 53, 114, 119, 188, 260
Kolobrzeg, Poland, 156
Krawczynski, Steve, SSgt., tail gunner, Lang crew, 96, 161, 163, 224-230, 258
Krems, Austria. See Stalag 17B
Kugelfisher Works, Schweinfurt, Germany, 144
Kutzberg, Germany, 177

LaGuardia Airport, New York, 229
Laimbach, Austria, 180
Lang crew, 94, 161, 258
Lang, Robert S., 2nd Lt., pilot, Lang crew, 30, 44-47, 68, 71, 93, 95-96, 98-99, 161, 163-164, 224-225, 229, 258, 263
Laon, France, 66
Larrick, Stanley H., TSgt., radio operator, Skerry crew, 99, 100, 103, 167, 173, 259, 263
Lawrence, Manning L., 2nd Lt., navigator, Bullock crew, 188, 260
LeFebre, George G., SSgt., tail gunner, Fisher crew, 94, 156, 258
Legg, Bret, Brig. Gen., US Military Attache, Switzerland, 204, 206, 209, 210
LeHavre, France, 185
Leipzig, Germany, 252
Lenning, Floyd A., SSgt., right waist, Holt crew, 86-88, 153-154, 257
Lepore, Domonic C., SSgt., tail gunner, McDarby crew, 76, 148-149, 256
Letanosky, Mike S., SSgt., tail gunner, Holt crew, 86-88, 154, 257, 263
Levy, Lester J., Sgt., right waist, Murdock crew, 30, 61, 74, 145-147, 150, 255
Lewis, Richard W., TSgt., right waist, Kenyon crew, 18, 56-57, 97-98, 165-166, 258
Leyan, von der, family, 169
Lille, France, 226
Limmel, Holland, 75, 145, 147
Lindquist, John C., 2nd Lt., copilot, Skerry crew, 100, 167-168, 259
Linrud, Arthur E., Sgt., top turret, McDarby crew, 53, 75-76, 148-149, 256
Linz, Austria, 181
Lithuania, 152
Lloyd, John W., Sgt., ball turret, Murdock crew, 74, 146-147, 150, 255
Lobendorf, Austria, 179
Loburg, Germany, 159
Loë, Count von, 169
Loë, Felix Freiherr von, 169
Loë, Therese Freifrau von, 169, 173-174
Long range belly tanks (P-47), 63
Loverich, Germany, 163
Lozenski, Charles B., SSgt., left waist, Kane crew, 42, 254
Luftgaukommando VI, 167
Luftwaffe, 15, 17, 58, 109, 137, 139, 145. See German Air Force
Luijten, Peter H., 146
Luke, H.W., 2nd Lt., tail gunner, Kane crew, 42, 254
Lutz, William E., SSgt., ball turret, Reid crew, 55

Maas river, 80, 151
Maastrich, Holland, 75, 145-147, 162, 232-233

Madrid, Spain, 224, 227, 244
Maercia, John, 228
Magdeburg, Germany, 159
Malmedy, Belgium, 67
Malmedy-Koblenz region, 67-68
Manahan, John C., 2nd Lt., navigator, Murdock crew, 75, 146, 255, 261-262
Manley, Harvey A., 2nd Lt., bombardier, McDarby crew, 76, 148-149, 256, 261
Marienburg, Germany, 143
Marion, Col., 1st Division Staff, 136, 142-143
Marionbrunn, Germany, 175
Martin, Bernard T., SSgt., radio operator, Kincaid crew, 178, 259, 265
Martin, William J., 2nd Lt., navigator, McDarby crew, 76, 148-149, 256, 261
Mathy, Major, Swiss Commandant, 207-208
Mauritania, Africa, 245
Mauthausen, Austria, 181
Maxwell crew, 107, 175, 259
Maxwell, Victor C., 2nd Lt., pilot, Maxwell crew, 66-67, 105, 107, 108, 109, 114, 175, 176, 187, 259, 263
Maynard, Kenneth A. SSgt., ball turret, Lang crew, 96, 162-163, 258
Memal (Klaipeda), Lithuania, 156
Menzies, William B., SSgt., tail gunner, Murdock crew, 74, 146-147, 255
Mercer, Finley J., Jr., TSgt., top turret, Kenyon crew, 98, 166, 258
Metcalf, Robert D., 2nd Lt., navigator, Kincaid crew, 112-113, 177-178, 259
Metz, France, 124
Miami, Florida, 245
Middleby, Robert J., SSgt., top turret, Skerry crew, 100, 168, 259
Midland, Texas, 191
Miller, John E., SSgt., togglier, Murdock crew, 74-75, 146-147, 255, 261
Milton, Theodore, Lt. Col., 48, 53, 59-61, 63-64, 67, 70, 73, 98-99, 106, 109, 111, 115-117, 123-125, 127, 136, 138, 141, 153, 175, 176, 252; comments at 1st Division critique, 137; 1st Combat Wing and 91st Bomb Group air commander, 59; flies adjacent to Cologne and passes over Bonn, 83-85; leads 1st Division (-) off course, 70; returns to course, 105; usurps command of 1st Division, 59;
Mission #115, 139. See Schweinfurt, second raid
Mockern, Germany, 159
Molen, Herman E., SSgt., togglier, Eakle crew, 30, 42, 77-78, 150, 213-216, 256
Morocco (Marrakech), Africa, 245
Munich, Germany, 124
Münsingen, Switzerland, 210
Munster, Germany, 143
Murdock crew, 73, 145, 255
Murdock, Douglas, L., 2nd Lt., pilot, Murdock crew, 30, 46, 47, 61, 73-75, 89, 103, 145-147, 150, 221, 255, 261-262
Mussolini, 207

McConnell, Warren E., TSgt., radio operator, Lang crew, 96, 162-163, 224, 258
McDarby crew, 75, 147, 255
McDarby, Dennis J., 2nd Lt., pilot, McDarby crew, 47, 53, 75-76, 147-149, 255

McGehee, Thomas K., Lt. Col., CO, 305th Bomb Group, 35, 50, 129, 134, 141, 251

Nantes, France, 42
Nazis, 225, 227-228, 237
Nemunas river, Lithuania, 156
Neubrandenburg, Germany, 157
Niemiec, Thaddues, J., SSgt., tail gunner, Farrell crew, 107, 124, 254
Norden bombsight, 23, 110-111, 122
Normand, Charles G.Y., Major, 44, 46, 47-48, 55, 63, 66-67, 70, 73, 75, 79, 85, 93, 98-99, 103, 105-106, 107, 108, 109-111, 114-116, 117, 135, 137, 145, 153, 167, 175-176, 187, 191, 251, 254; aircraft commander and group leader, 33-34; arrives Orfordness late, 54; arrives Spalding early, 52; CO, 365th Bomb Squadron, 33-34; awarded Silver Star, 252; comments at 1st Division critique, 140-142; departs English coast alone, 54; departs Thurleigh late, 49; departs Spalding early, 52; fails to reach Daventry, 50; flying experience, 34-35; group leader, 34; maneuvers over channel; 60-62; occupies position of 381st Bomb Group, 64; promoted to Lieutenant Colonel, 252; recommends himself for a decoration, 134; shot down, 252
Norris, Donald H., TSgt., radio operator, Eakle crew, 78, 149-150, 256
North Africa, 207
North African Campaign, 149

Oldenburg, Germany, 68
Orfordness, England, 48-49, 52-54, 55, 58, 136, 140-141
Oxford General Hospital, England, 192

Pacific Theater of Operations, 42, 112
Palace (Nevada) Hotel, Switzerland, 206, 210
Palmen, Herr Josef, 163-164
Paris, France, 66, 224-227, 238-240
Paroles, 207-209
Pearl Harbor, 42, 228
Peaslee, Budd J., Col., 48, 50, 52-54, 55, 63, 70, 72, 83, 105, 106, 109, 117, 123, 127, 136-137, 141, 142, 252 comments at 1st Division critique, 138-139; 40th Combat Wing, 1st Division, and 92 Bomb Group air commander, 47; maneuvers over channel, 58-60; turns division lead over to 1st Combat Wing, 58-59
Pellegrini, Joseph, 1st Lt., bombardier, Kane crew, 107, 108, 109-111, 115-116, 118, 129, 135, 142, 254
Piarco Field, Trinidad, British West Indies, 34
Pikelis, LeRoy V., TSgt., top turret, Kane crew, 54, 60, 85, 254
Pilot's Directional Indicator (PDI), 110, 142
Polish pilots, 43
Protestant Church of Adelboden, Switzerland, 210
Puerto Rico, 245
Puffendorf, Germany, 97, 161, 163-164
Pumo, Jerome B., TSgt., radio operator, Maxwell crew, 108, 176, 259, 265
Purple Heart, 30
Purple Heart corner, 46, 73
Puth, Holland, 230
Pyrenees mountains, 224, 227, 241

Quackenbrück, Germany, 63, 117

Raines, John F., SSgt., top turret, Kincaid crew, 30, 108,

Index

112-113, 177-178, 259
Red Cross, American, 185; International; 206, 217
Regensburg, Germany, 16-17. See Schweinfurt
Reid, R.E., 2nd Lt., pilot, Reid crew, 45, 55, 67, 106, 118, 175
Rheims, France, 33, 125
Rhine river, 93, 98, 100, 103, 196
Roberts, Benjamin F., SSgt., ball turret, McDarby crew, 76, 148-149, 256
Rodersdorf, Switzerland, 199
Roe, G.M., SSgt., right waist, Kane crew, 129, 254
Roermond, Holland, 80-81, 151
Rohr, Lt. Col., 63, 70, 83, 105, 109, 117, 123; 41st Combat Wing and 379th Bomb Group air commander, 59
Rollo, Frank W., SSgt., left waist, Holt crew, 86-88, 153-154, 257
Rommel, Erwin, Field Marshall, 226
Roosevelt, Franklin Delano, 180-181, 217
Rossbach, Austria, 183, 185
Roth, Willi, Oberfeldwebel (1st Sgt), 99, 100, 167, 170
Rowan, Willis V., 2nd Lt., copilot, Maxwell crew, 176, 259, 263
Rowlett, Wayne D., SSgt, ball turret, Skerry crew, 100, 103, 168, 259
Rowley, Donald T., 2nd Lt., navigator, Dienhart crew, 89, 196-200, 202-203, 257, 263; awarded Silver Star, 205; funeral of, 204-205
Roxendorf, Austria, 179
Royal Air Force, 15, 42, 178, 239; Battle of Britain, 18
Ruhr valley, 36

Saarbrücken, Germany, 199
Saguia el Hamra, West Africa, 245
Saint Florian, 174
Saint Nazaire, France, 57
Sanchez, Robert L., SSgt., ball turret, Eakle crew, 78, 149-150, 256
San Francisco, California, 164
San Juan, Puerto Rico, 245
San Sebastian, Spain, 243-244
Schinnen, Holland, 231
Schlatthof farm, Switzerland, 202
Schlippenback, Rainer Freiherr von, 168
Schouwen Island, Holland, 55, 65
Schranz, Hilda, 210
Schranz, Peter, 210
Schutzstaffel (SS), 181
Schweinfurt, Germany, 18, 31, 104, 106, 108, 109, 114-115, 119, 170, 175-178, 187, 192; ball bearing industrial losses, 144; bomb loading, 29; comments at 1st Division critique, 135-143; distance to England after bomb run, 159; estimated machine damage, 249; first raid together with Regensburg, 16-17, 36; second raid, 19
Sedinger, Edward J., SSgt., left waist, Willis crew, 80, 151-152, 216, 256, 263
Segal, Bernard, SSgt., tail gunner, Dienhart crew, 89, 91, 195-197, 203, 207, 209, 257
Seifert, Major, 74-75
Seine river, France, 239
Shackley, George, Maj., air commander 381st Bomb Group, 64
Shaull, Marvin D., TSgt., top turret, Farrell crew, 106-107, 254

Shuckanhauser Hotel, Switzerland, 204-205
Silver Star, 134, 205, 252
Sittard, Holland, 70, 151
Skerry crew, 99, 166, 258
Skerry, Robert A., 2nd Lt., pilot, Skerry crew, 47, 62-63, 79, 93, 98, 99, 100, 103, 161, 166-168, 169-170, 258
Smart, Jayson C., SSgt., right waist, Farrell crew, 43, 53, 72, 107, 129, 254
Smith, Barden G., TSgt., top turret, Eakle crew, 77-78, 150, 256
Smith, Edwin L., 2nd Lt., copilot, Murdock crew, 30, 46, 73-75, 146-148, 255
Smith, Hurley D., TSgt., radio operator, Dienhart crew, 90, 196-198, 202-203, 206, 211, 257
Smith, Norman W., 2nd Lt., copilot, Kincaid crew, 42, 111-112, 176, 178, 259, 265
Snow, Bernard E., SSgt., ball turret, Willis crew, 79, 152, 256
Southampton, England, 218
Soviet Union, 156
Spadafora, Alfredo A., SSgt., left waist, Eakle crew, 77-78, 89, 149-150, 256
Spain, 196, 221, 224, 226-227, 241-244
Spalding, England, 50, 52, 140-141
Spanish Sahara, West Africa, 245
Splasher #6, England, 52-53
Squadrons (numbered):
  350th Fighter Squadron, 65
  351st Fighter Squadron, 65
  352nd Fighter Squadron, 65
  364th Bomber Squadron, 29, 35, 44, 46, 104
  365th Bomber Squadron, 29, 33, 44
  366th Bomber Squadron, 29, 35, 44, 47, 103, 106
  422nd Bomber Squadron, 29
Stalag Luft 3, 147, 149, 151, 163, 166, 168, 252
Stalag Luft 4, 156
Stalag Luft 6, 156, 217
Stalag 7A, 166
Stalag 7B, 147
Stalag 9C, 217
Stalag 17B, 147, 149, 151, 154, 163, 166, 168, 176, 178, 188, 190, 213, 216, 218. See Krems, Austria
Stanchak, Nicholas, SSgt., tail gunner, Willis crew, 80, 151, 263, 256
Stendal, Germany, 157
Stony Stratford, England, 48-49
Stuttgart, Germany, 56, 124, 190
Swinemünde (Swinoujscie), 156-157, 217
Swiss Honor Guard, 205
Switzerland, 188-189, 196-200, 204, 206-207, 210, 251

Tablets of the Missing, 253
Task, Harry J., 1st Lt., pilot, Task crew, 47, 49
Tent City, 159. See Camp 11A
Tew, John C., Jr., 2nd Lt., navigator, Lang crew, 71, 95-96, 162-163, 258
Therrien, Thomas E., SSgt., right waist, Fisher crew, 94, 155, 258, 263
Third Reich, 152
Thurleigh, England, 35, 45-49, 52-53, 58, 67, 140-141
Tongeren, Belgium, 146, 221
Torgau, Germany, 159
Tourcoing, France, 226, 236, 238
Travis, Robert F., Brig. Gen., CO, 41st Combat Wing, 139

Turner, Howard M., Col., CO, 40th Combat Wing, 138, 142
Tuttlingen, Germany, 190

Uelzen, Germany, 157

Vancouver, Washington, 155
VE-Day, 193, 218
Vegesack, Germany, 17
Vereingte Kugellager Fabriken Works I and II, 144
Villacoublay, France, 42
VJ-Day, 193
Volkssturm, 179

Walcheren Island, Holland, 55, 65
Waldenrath, Germany, 94, 155
Weddings: England, 194, Switzerland, 209-210
Wells, Robert G., Sgt., right waist, McDarby crew, 76, 148, 256, 261
Wertheim, Germany, 105, 124
West Reynham, England, 49
West Schouwen, Holland, 65
White Cliffs of Dover, 61
Whitehead, Herbert S, SSgt., left waist, Maxwell crew, 108, 175-176, 259
Wiesenburg, Germany, 159
Wiggens, Thelma B., Jr., TSgt, radio operator, Murdock crew, 74, 146, 221, 255
Wilhering, Austria, 181
Williams, Robert B., Brig. Gen., C0, 1st Bombardment Division, 135, 137-140, 142-143
Willis crew, 79, 151, 256

Willis, Charles W., Jr., 2nd Lt., Willis crew, 45, 47, 62, 79-81, 98, 99, 151, 216, 256, 261
Wilson, Delmar E., Col., previous CO, 305th Bomb Group, 35
Wilson, Lloyd G., SSgt., tail gunner, Eakle crew, 77-79, 150-151, 221-224, 256
Wings (combat box):
1st Combat Wing, 48-50, 53-54, 59-61, 63-64, 65-67, 70, 79, 83, 85, 99, 105, 106, 109, 111, 116-117, 119, 121, 123-124, 128, 136-142, 153, 175; composition of, 31;losses, 248
40th Combat Wing, 31, 47-50, 52-53, 54, 55, 58-61, 63, 65-67, 70, 83, 85, 105, 109, 111, 123, 136-139, 142, 252; composition of, 31; losses, 248
41st Combat Wing, 52, 53-54, 59, 63, 65-66, 70, 83, 85, 105, 109, 111, 117-119, 123, 138-139; composition of, 31; losses, 248
Würzburg, 105, 106, 107, 108, 175

Young, Aubrey C., Jr., 2nd Lt., copilot, Holt crew, 86-87, 154, 257, 263
Ysperdorf, Austria, 180
Yugoslavia, 215
Yugoslavian partisans, 215

Zavar, Andrew J., 2nd Lt., bombardier, Maxwell crew, 108, 176, 259, 265
Zeeland, Holland, 31, 55
Zeuren, Leo, 80
Zullo, Christy, SSgt., left waist, Dienhart crew, 43, 89-90, 197, 198, 203, 209-210, 257